AF702404

The Microphone Handbook

The Microphone Handbook

John Eargle
First Edition

Elar Publishing
Plainview, New York

LIBRARY OF CONGRESS
CATALOG CARD NO: 81-70852

Eargle, John
The microphone handbook.

Plainview, NY: Elar Publishing Co.
256 p.
8112 811023

All rights reserved. No part of this book may be reproduced in any form or by any means without permission in writing from the publisher.

Illustrations by Hal Keith
Page layout by K&S Graphics, Island Park, N.Y.

Table of Contents

Foreword

I first met John Eargle in 1965, at the old RCA studios on 24th Street in New York City. I had just joined RCA as Musical Director in the classical Red Seal division. John was director of Quality Control at the time. As it happened, we both were engaged in special projects which required night-work. It didn't take long to establish a series of nightly "bull-sessions" (in which we quite naturally solved all of the problems in audio, as well as those that beset the world in general). Those "bull-sessions" were also a very revealing insight into this multi-talented, very complex man that is John Eargle.

John Eargle is one of those rare individuals–some call them "renaissance men"–who combine a profound knowledge of the theoretical and practical aspects of physics, acoustics, electronics and audio engineering, along with the intellectual capacities and disciplines which are manifested in artistic expression. John has a lifelong commitment to music. He is an accomplished pianist and particularly enjoys playing his recently acquired Bösendorfer grand piano.

In addition to holding a Master's Degree in Music from the University of Michigan, John also has obtained a Master's Degree in Engineering from Cooper Union.

This book is about microphones and their use. Is it necessary to have degrees in music to understand microphones? Probably not. But certainly in the area of applications and the primary use of microphones–the recording of music–a knowledge of the music, and the physics of music, can be invaluable. John Eargle combines this knowledge with his experience as former Chief Engineer of Mercury Records. John quite rightly eschews the common practice of choosing microphones by some sort of esoteric subjective "mumbo-jumbo." John Eargle believes in a thorough, meticulous, scientific explanation of the operating principles of the various types of microphones. After establishing these facts, John offers a scientific rationale for his selection of microphones which will provide optimum results in a specific application.

Those looking for a "cookbook" approach to microphone usage will not find it here. Recording engineers, and all others who require an accurate and comprehensive reference guide to microphones and their applications, will find this book invaluable.

Bert Whyte
Recording Engineer
Associate Editor Audio Magazine
New York

Author's Preface

The microphone is one of the most important tools in the disciplines of recording, broadcasting and sound reinforcement, and a thorough understanding of its operation and use is essential for a student or apprentice aspiring to a career in one of these fields. Not too many years ago, the beginning engineer had to learn his craft on the job. It was easiest for him simply to emulate an experienced engineer without taking the time required to learn the physical principles of how microphones really work. A person who learns his trade in this way may become a competent, but not flexible, engineer, and come to rely too heavily on certain microphone models, and much of his problem solving may be by rote.

It is the author's conviction that the student who learns principles also acquires analytical skills. With the principles in place, the particulars of a given problem can easily be solved. The student who knows "why," as well as "how," obviously can progress rapidly in his chosen field. This is the premise for *The Microphone Handbook*. It teaches fundamental principles; problems are solved through analysis, and the student is encouraged to think of basic microphone types rather than specific models.

The last ten years have seen many colleges develop four-year programs in the recording arts and sciences. Other schools offer summer workshops in this field, and many competent recording engineers have set up their own studio-oriented educational activities. This book is intended for use at both the collegiate level and the more informal workshop level. The treatment is essentially non-mathematical, and the emphasis is on graphic methods of presentation. It is desirable, but not essential, that the student using this book have a familiarity with acoustics and a broad knowledge of electro-acoustical systems.

The book is broadly divided into four sections. The first presents fundamentals of how microphones operate and covers the following:

1. Basic Transducer Theory
2. Directional Characteristics: The Basic Patterns
3. Microphone Sensitivity Ratings
4. Remote Powering of Capacitor Microphones
5. Microphone Care and Maintenance

The next section deals with the microphone in its physical environment and covers:

6. Using Patterns Effectively
7. Proximity and Distance Effects
8. Reflections from Nearby Surfaces
9. Multi-microphone Interference Problems

The third section treats the microphone in its various aesthetic environments:

10. Stereo Microphone Techniques
11. Speech and Music Reinforcement
12. Studio Microphone Techniques

The last section deals with two practical topics which are not always given the attention they deserve:

13. Electrical Interface Problems
14. Microphone Accessories

A bibliography is included for those students who wish to broaden their knowledge in the allied arts as well as delve more deeply into the technology of microphone design.

The author is grateful to the following manufacturers who have supplied photographs and drawings for use in the book: Altec Corporation, Atlas Sound, Audio-Technica, Bruel & Kjaer, Crown International, Edcor, Electro-Voice, Gotham Audio Corporation, Journal of the Audio Engineering Society, JVC, G. Neumann GMBH, Sennheiser, Schoeps, and Shure Brothers.

CHAPTER I

Basic Transducer Theory

INTRODUCTION

A transducer is a device which transfers energy from one system to another, usually in a different form. The microphone is one of the earliest *electro-acoustical* transducers. Its origins go back about a century, to the introduction of telephony. But it was not until the late twenties that the microphone was to be used for sound recording. During the preceding decade, the microphone had been introduced to the emerging fields of sound reinforcement and radio broadcasting.

In the United States, Western Electric and RCA played major roles in the early development of microphones, followed in later years by Electro-Voice and Shure Brothers. In Europe, Neumann in Germany and AKG in Austria have been important contributors.

While just about every known means of acoustical-to-electrical transduction has been used in microphone design, the present art has settled on magnetic and capacitor types, which are inherently capable of flat frequency response. The carbon-granule microphone was developed for telephony, and is still used for that purpose. Crystal microphones, using the piezo-electric principle of operation, were at one time used extensively for paging applications, but they have been largely replaced by moving-coil types.

Microphones have improved dramatically and become lower in price over the years, and many specialized types have been developed for particular applications and use under unusual environmental conditions. In this chapter, we will examine the four basic types of transduction listed above.

THE CARBON MICROPHONE

The carbon microphone operates on the principle of varying the resistance of loosely-packed carbon granules as they react under the varying pressure of sound waves. The basic design is shown in Figure 1-1.

Figure 1-1. Details of a carbon microphone. **(A)** Cutaway view; **(B)** circuit diagram; **(C)** typical response.

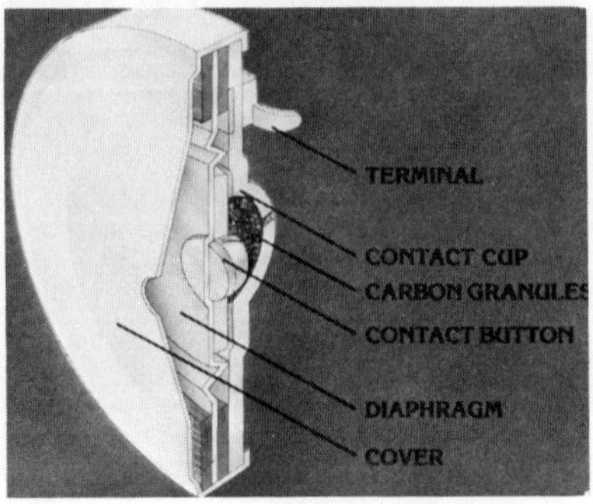

(A)

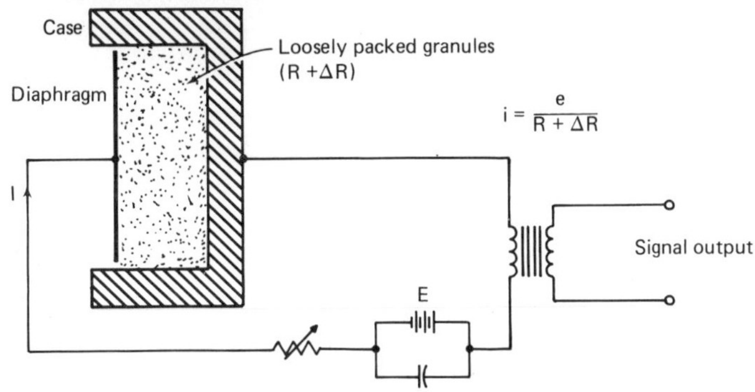

(B)

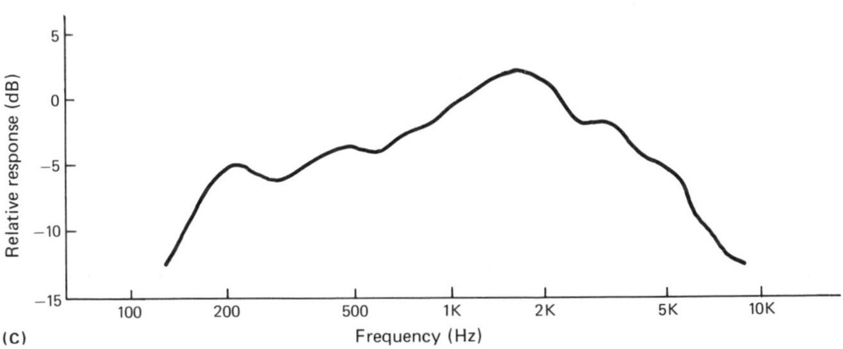

(C)

When no signal is present, the resistance, R, of the carbon microphone is constant. When sound waves impinge on the diaphragm, the resulting vibrations cause a corresponding alternating change in R. As the diaphragm moves inward, compressing the carbon granules, R decreases: as the diaphragm moves outward, R increases. The change in R, symbolized as ΔR, causes a corresponding change in the current flowing through the carbon element. The Greek letter, Δ *("Delta"),* represents a small change in a quantity. The change can be either positive or negative, and—in the case of audio signals—will vary with time. The varying current can modulate an earphone directly, as in early telephony, or it can be amplified as required.

The carbon microphone is very limited in dynamic range and frequency response, and it exhibits high distortion. It is rarely used outside of telephony, where its performance characteristics and low cost work to its advantage.

THE PIEZO-ELECTRIC MICROPHONE

Some crystalline substances, if bent or twisted along certain of their axes, will generate an electrical potential between opposite faces. The effect is known as *piezo-electricity* (from the Greek *piezein* meaning, "to press"). The crystal can be coupled to a diaphragm and the device used as a microphone. While the performance characteristics of the crystal are better than those of the carbon microphone, they are fairly limited, relegating crystal microphones to no more than paging applications. Details of the crystal microphone are shown in Figure 1-2.

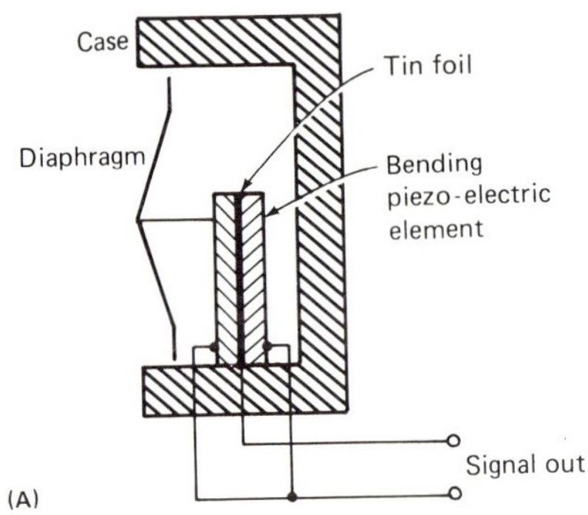

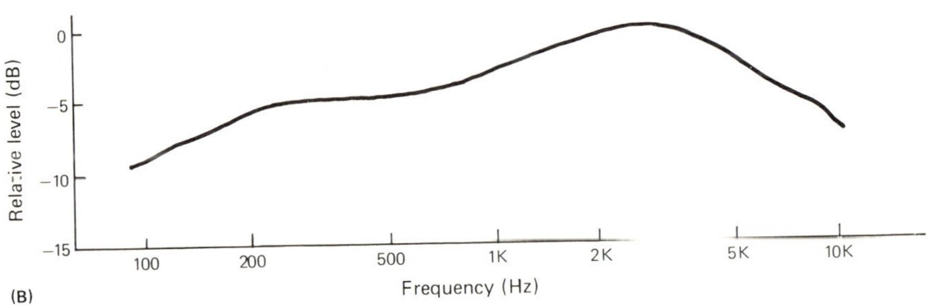

Figure 1-2. Details of a piezo-electric microphone. **(A)** Circuit diagram; **(B)** typical response.

DYNAMIC MICROPHONES

The Moving-coil Microphone

The principle of magnetic induction underlies the operation of the dynamic microphone: if a conductor is moved within a magnetic field, a voltage is induced across the conductor, as shown in the equation

$$e = Blv$$

where:

e = the induced potential, in volts;
B = the magnetic flux density, in teslas;
l = the conductor length, in meters;
v = the velocity of movement, in meters-per-second.

In its simplest form, the dynamic moving-coil microphone consists of a fine coil of wire, suspended in a magnetic gap. The coil is attached to a dome-shaped diaphragm which acts as a receiver for sound waves. Details are shown in Figure 1-3.

Modern dynamic moving-coil microphones are quite small, and some may have diaphragms no more than 1.5 cm (0.6 in) in diameter. Coils of very fine wire with many turns acting in high-energy magnetic gaps result in relatively high output levels. Attention to diaphragm suspension design results in linear, low-distortion output, and careful damping of cavities in the neighborhood of the diaphragm results in smooth, peak-free response. Current models are extremely rugged.

For flat output over a wide frequency range, the diaphragm of a dynamic microphone must execute constant velocity with respect to frequency when it is placed in a sound field of constant pressure with respect to frequency. In order to ensure this, the diaphragm must be well damped, so that it is *resistance-controlled.* This will be discussed in considerable detail in Chapter 2.

The Ribbon Microphone

The ribbon microphone also operates on the principle of magnetic induction. A corrugated ribbon, suspended in a magnetic gap, develops a signal voltage as it moves in response to sound waves. Because of the relatively-short conductor length (as compared with the multi-

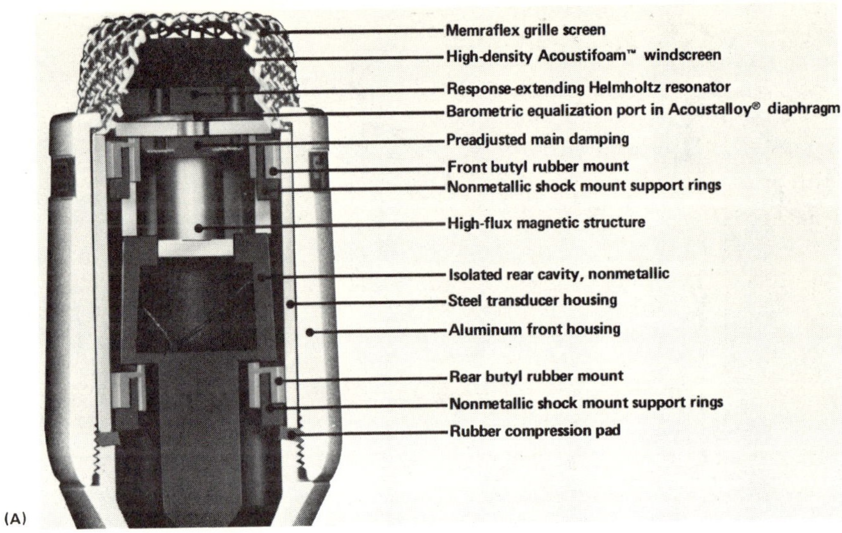

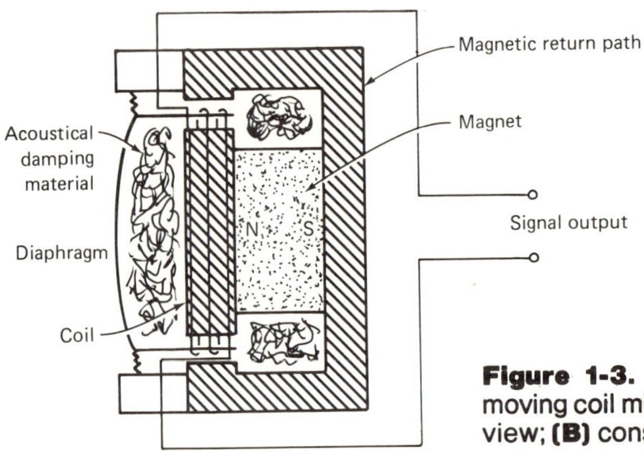

Figure 1-3. Details of a dynamic moving coil microphone. **(A)** Cutaway view; **(B)** construction details.

turn coil described above), and the large magnetic gap with its relatively low magnetic field strength, the output of the ribbon is very low. It is customary to use a step-up transformer in the design of the microphone to attain a useful output level. Details of the ribbon microphone are shown in Figure 1-4.

The ribbon microphone is also known as a *velocity*, or *pressure gradient*, microphone; that is, it responds to the air particle velocity which is developed by the pressure gradient (i.e., difference) between the two sides of the ribbon. Side- and top-view details are given in Figure 1-4. In the top view, note the difference in path lengths, P_1 and P_2, for a sound arriving at an angle perpendicular to the ribbon. For

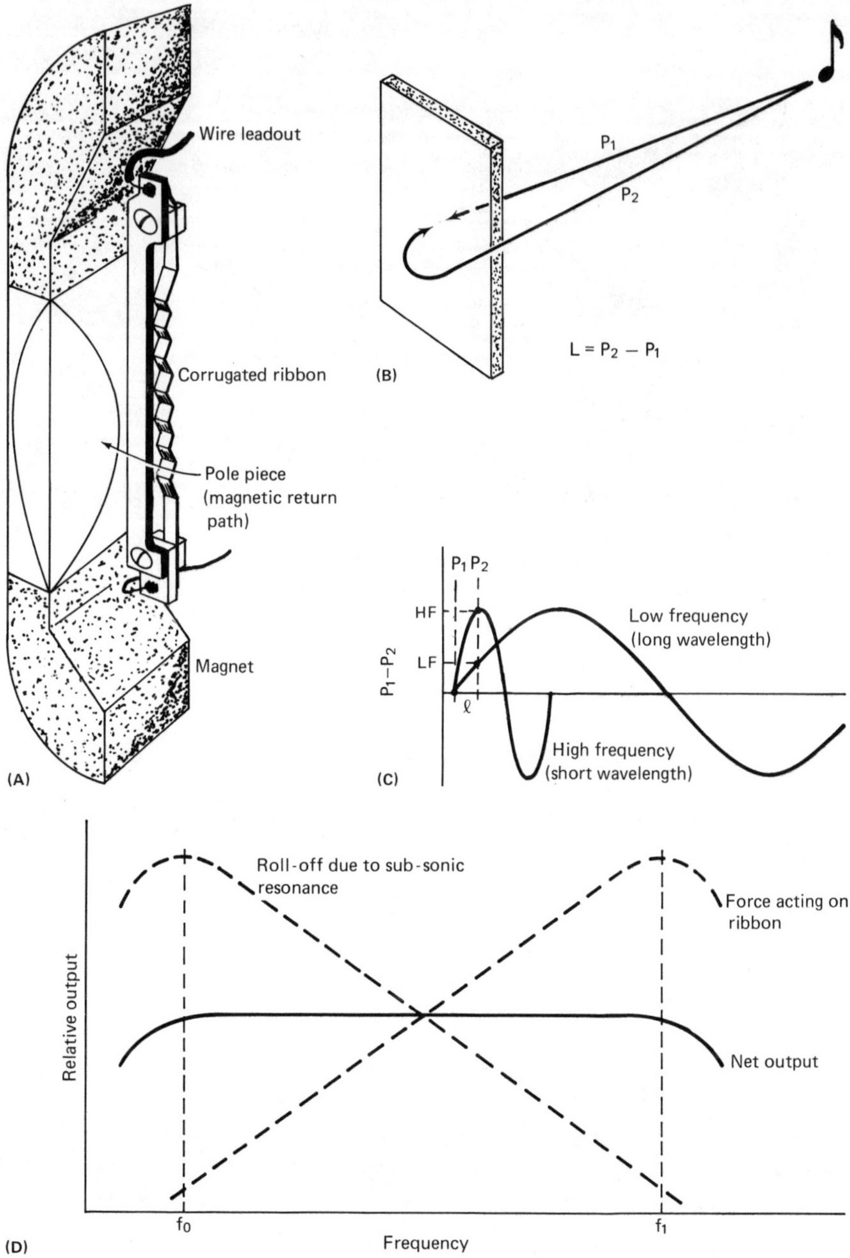

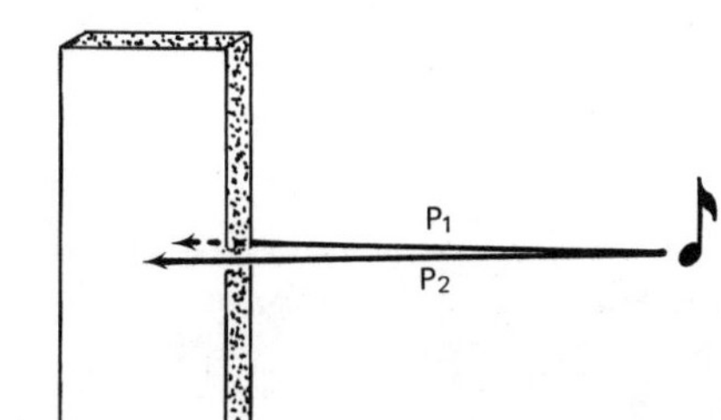

Figure 1-4. Details of a dynamic ribbon microphone. **(A)** Cutaway view; **(B)** top view. **(C)** Note that the pressure difference, $P_1 - P_2$, increases with rising frequency. **(D)** The effect of force and roll-off on output. **(E)** side view.

low frequencies, with their relatively long wavelengths, the path length difference corresponds to a small portion of the wavelength, and the pressure difference ($P_1 - P_2$) is fairly small, as shown in Figure 1-4C. In the figure, we may also note the action of a high frequency, with its shorter wavelength. Here, the pressure difference is larger, and the driving force on the ribbon is correspondingly greater. The force on the ribbon rises with frequency at a rate of 6 dB-per-octave. But, because of its sub-sonic resonance frequency, f_0, the ribbon itself tends to roll off with rising frequency, also at the rate of 6 dB-per-octave. As shown in Figure 1-4D, these two characteristics are complementary, and the net effect is a flat response. If the ribbon is made small enough, its response extends well up into the 15-to-20 kHz range. The upper limit, f_1, of the ribbon microphone's response is determined by interference effects at short wavelengths (high frequencies) around the pole pieces. Specificallv, for useable response out the 15 kHz, L would have to be such that it equaled the wavelength (λ) for sound in air, of a frequency of about 30 kHz. That is:

$$\begin{aligned} L = \lambda &= c/f \text{ (speed of sound/frequency)} \\ &= \frac{344 \text{ meters-per-second}}{30 \text{ kHz}} \\ &= 1.15 \text{ cm } (0.4 \text{ in}) \end{aligned}$$

Finally, in Figure 1-4E, note that for signals arriving from the side, the equal path lengths, P_1 and P_2, result in no pressure difference across the ribbon, and thus there is no output. For this reason, the typical ribbon microphone design exhibits a bi-directional, or "figure-8", response, although modifications in construction details may allow other response patterns as well. These will be discussed in greater detail in the following chapter.

THE CAPACITOR (CONDENSER) MICROPHONE

The capacitor is a circuit element which stores a charge when an electric potential is impressed across it. The equation relating these quantities is

$$Q = CV$$

where:

Q = the charge, in coulombs
C = capacitance, in farads
V = potential, in volts.

In the capacitor microphone, the plates of the capacitor are designed so that one of them (the back plate) is stationary, while the other plate is a thin, taut metallized plastic diaphragm. When sound waves impinge on the diaphragm, it moves back-and-forth, relative to the fixed back plate. The resulting variation in capacitance, ΔC, operates to produce a corresponding variation of potential across the capacitor, according to the equation

$$V + \Delta V = \frac{Q}{C + \Delta C}.$$

For flat output over a wide frequency range, the diaphragm of the capacitor mircophone must execute *constant displacement* with respect to frequency, when placed in a sound field of constant pressure with respect to frequency. This requires that the diaphragm be taut enough so that its primary resonance lies well beyond the intended frequency range of the device, and that it operate over its *stiffness-controlled* region. This will be discussed in more detail in Chapter 2.

Details of the capacitor microphone are seen in Figure 1-5, where we show the effect of altering the spacing between the plates of a capacitor with a fixed charge. We also see a typical circuit for a capacitor microphone. The voltage source, shown here as a battery, polarizes the capacitor through the resistor, R, of high enough value (typically, well into the megohm range) so that the charge on the capacitor remains fixed down to the lowest frequency at which the microphone is expected to respond. The variation in signal potential across the capacitor is quite small, and it is invariably amplified adjacent to the capacitor element itself, as shown in the figure.

Modern capacitor microphones are quite stable, and can be built to very small dimensions. Accordingly, they can exhibit extremely wide frequency response. For these reasons, they have become standard for precision sound pressure measurements.

Electret Microphones

In recent years, the *electret* principle has been used in capacitor microphone design. An electret is a material capable of holding a charge with no need for an external polarizing voltage. Both waxes and fluorocarbons have been used for this purpose. Early problems of instability have been overcome, and current electret microphones are characterized by excellent frequency response and long-term stability. Details of the electret principle are shown in Figure 1-6. In

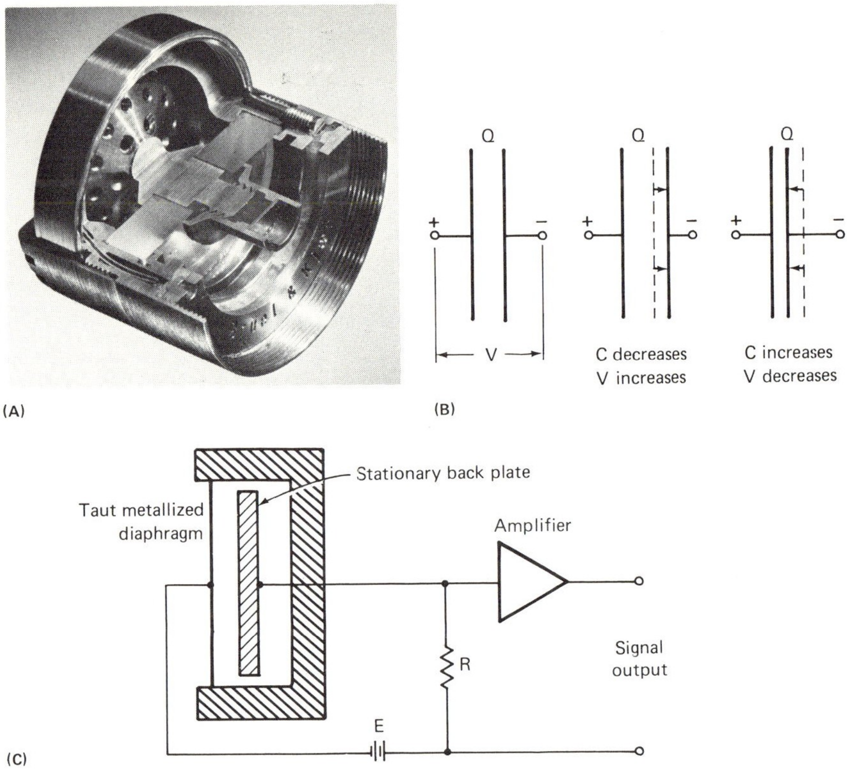

Figure 1-5. Details of a capacitor microphone. **(A)** Cutaway view, showing perforated backplate *(Bruel & Kjaer photo)*. **(B)** Capacitance and potential relationships with fixed voltage. **(C)** circuit diagram.

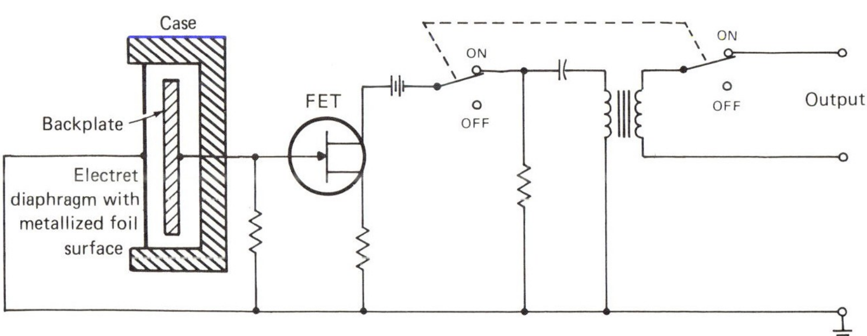

Figure 1-6. Details of an electret microphone, showing the electret element and a typical circuit diagram.

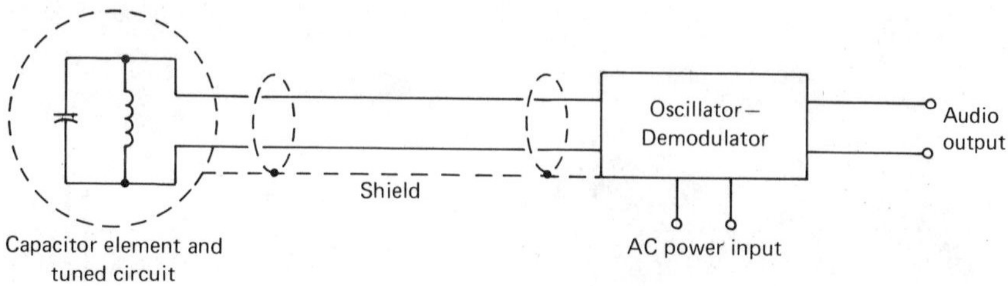

Figure 1-7. Block diagram of an RF capacitor microphone.

the design shown in the figure, the electret is in the form of a foil, and is integral with the diaphragm. The so-called *back-electret* design uses a conventional metallized diaphragm with a stationary electret integral with the back plate.

RF Capacitor Microphones

Another kind of capacitor microphone, rarely encountered these days, operates on RF (radio frequency) principles, without a polarizing voltage. A typical block diagram is shown in Figure 1-7. In this design, the capacitor element is part of a tuned circuit. Minute changes in capacitance alter the tuning of the circuit, producing a frequency-modulated signal in much the same manner as FM signals are produced in broadcasting. These signals are then demodulated in the conventional way, and useful audio output is obtained.

At one time, RF capacitor microphones enjoyed an advantage in terms of sensitivity and extended frequency response, due to their relatively small size. They tended however to be noisy and somewhat unstable. They were more susceptible to certain kinds of RF interference. With the many improvements made in conventional and electret design, they have pretty much passed from the scene.

CHAPTER II

Directional Characteristics: The Basic Patterns

INTRODUCTION

One of the most important characteristics of a microphone is its *directionality:* the way it responds to sounds arriving from different directions. The engineer normally has at his disposal microphones of various directional characteristics, and he will choose among them, depending on the acoustical environment and the program requirements at hand.

There are three essential patterns: *omni-, bi-,* and *uni-directional.* Two of these (omni- and bi-directional) may be used to produce several variations of the uni-directional pattern. The uni-directional pattern can also be produced with a single-diaphragm microphone, by precisely controlling the sound entering through paths in the microphone case to the rear of the diaphragm. In fact, most uni-directional microphones encountered today are of the single-diaphragm type.

All of the methods used for attaining specific directional characteristics have been known for years. As we review these, we will relate them to significant early microphone models which embodied these methods.

MEASUREMENT OF DIRECTIONAL CHARACTERISTICS

Directional response patterns for microphones are invariably shown in *polar coordinates,* and the resulting graph is called a *polar plot.* In making these measurements, a microphone is placed on a turntable capable of 360-degree rotation. The microphone's diaphragm is placed precisely at the center of the turntable. A loudspeaker is placed at some distance from the microphone, and is fed a steady signal at a fixed frequency. As the microphone is rotated, a circular chart on a graphic level recorder moves in synchronism with it. The moving pen of the graphic level recorder registers the output level from the

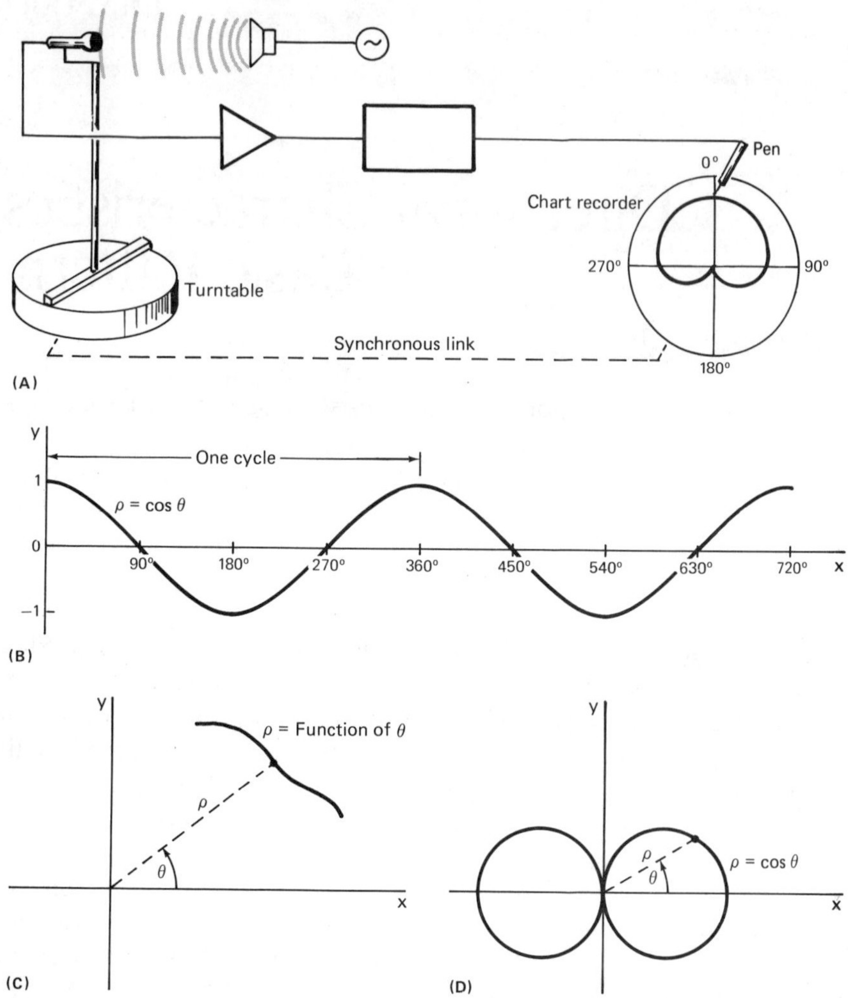

Figure 2-1. Apparatus for recording polar response of a microphone. **(B)** The function, $\rho = \cos\theta$, plotted in rectangular coordinates. **(C)** A general polar function **(D)** plotted in polar coordinates.

microphone, either in voltage, or in decibels relative to a fixed reference level. Details of this procedure are shown in Figure 2-1.

Polar coordinates are useful for plotting any function which has a cyclic, or repeating, angular pattern. Most readers are of course familiar with *rectangular* coordinates, as shown in Figure 2-1B. Here, we have plotted the function, Y = cosX. Note that the pattern repeats itself every 360 degrees. In polar notation, the independent variable is expressed directly as the angle, θ (Greek *"Theta"*), while the dependent variable is expressed as the distance, ρ (Greek *"Rho"*), measured from the center of the graph, as shown in Figure 2-1C. When the function, $\rho = \cos\theta$ is plotted, we have the form shown in Figure 2-1D. Normally, we consider only values of θ between 0-and-360 degrees, since the pattern merely repeats itself for multiples of 360 degrees. The use of polar coordinates in illustrating the angular response of microphones is intuitively obvious; θ is simply the bearing angle around the microphone, and ρ is the microphone's relative output in that direction.

Figure 2-2 shows polar plots of output voltage and output level in dB for the same microphone. The two curves convey essentially the same information. The form shown in Figure 2-2B is typical of most manufacturer's published data.

Polar Equations for First-order Microphones

A *polar equation* may be written to describe the amplitude of any polar response, at any angle, θ. The equation is of the general form given here;

$$\rho = A + B\cos\theta \text{ (where } A + B = 1\text{)}.$$

The specific values of A and B define the actual shape of the polar

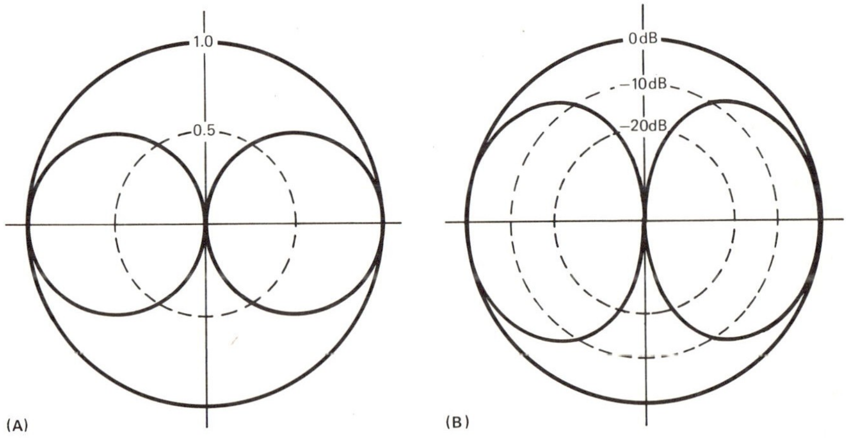

Figure 2-2. Representative polar graphs of **(A)** output voltage or sensitivity, and **(B)** output level, in dB.

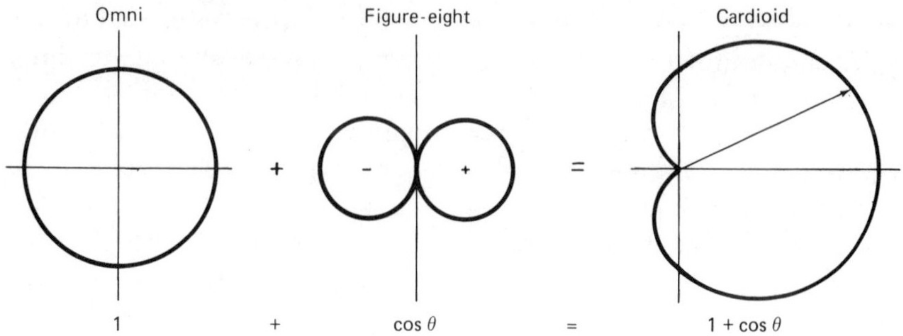

Figure 2-3. Construction of the cardioid pattern, resulting from the summation of omni and figure-8 patterns.

pattern. For example, when A = 1 (and therefore, B = 0), the pattern is omni-directional. When A = 0 (and B = 1), the pattern is bi-directional. To illustrate this, simply substitute these values in the equation given above. For example

$$\rho = 1 + 0\cos\theta = 1; \text{ omni-directional (pressure-only)}$$

and

$$\rho = 0 + 1\cos\theta = \cos\theta; \text{ bi-directional (velocity-only).}$$

From the above, we see that the omni-directional pattern must be constant in all directions, since the cosθ element is always zero. This is to be expected, since sound pressure alone conveys no information about angular direction, and a pressure-only device will simply read that pressure at a point in space, regardless of its orientation.

However, information about angular direction *is* contained in the air particle velocity, and this is the information to which the bi-directional pattern responds. Such microphones are sometimes called *cosine* microphones, as may be evident from the equation above. Note that the bi-directional pattern can only detect the axis along which a sound arrives. However, for any output level, that sound may be arriving from either the angle θ, or, θ ±180 degrees.

The Cardioid Pattern

When a velocity (cosine) element is used in conjunction with a pressure element, a uni-directional pattern is derived, and the directional ambiguity is resolved. This may be demonstrated by adding the polar equations for the omni-directional (pressure) and bi-directional (velocity) elements

$$\begin{array}{rl} 1 + 0\cos\theta & \text{omni-directional} \\ + \underline{0 + 1\cos\theta} & \text{bi-directional} \\ 1 + 1\cos\theta & \text{uni-directional} \end{array}$$

Often, polar equations that have been derived in this manner are "normalized" so that, as before, A + B = 1. This gives us a uni-directional polar equation of, $\rho = 0.5 + 0.5\cos\theta$. Note that, at 0 degrees, $\rho = 1$ for *any* normalized polar equation.

DUAL-ELEMENT CONFIGURATIONS

One of the earliest directional microphones was the Western Electric 639 design. This microphone consisted of a dynamic pressure unit and a ribbon velocity element, both located in the same housing, as seen in Figure 2-4. For sound sources along the frontal axis of the microphone, the pressure and ribbon elements add; at 180 degrees, the responses of the pressure and ribbon elements cancel, and the response of the microphone is effectively zero. At angles of ±90 degrees, the output is due only to the pressure element.

The dual-element capacitor microphone, also known as the Braunmühl-Weber configuration (after two pioneer designers, c. 1935), also embodies the characteristics of both pressure (omni-directional) and velocity (figure-8, or cosine) elements in a single structure. In order to show how it produces directional characteristics, we will first present a set of mechanical and electrical analogies:

Electrical		Mechanical	(Impedance analogy)	
Voltage	(E)	Force	(F)	
Current	(I)	Velocity	(V)	
Resistance	(R)	Damping	(D)	(dash pot)
Capacitance	(C)	Compliance	(C)	(spring)
Inductance	(L)	Mass	(M)	

When a mass is suspended by a compliance, or a spring, it will bob up and down at a natural frequency, f_0, given by

$$f_0 = \frac{\sqrt{1/MC}}{2\pi}$$

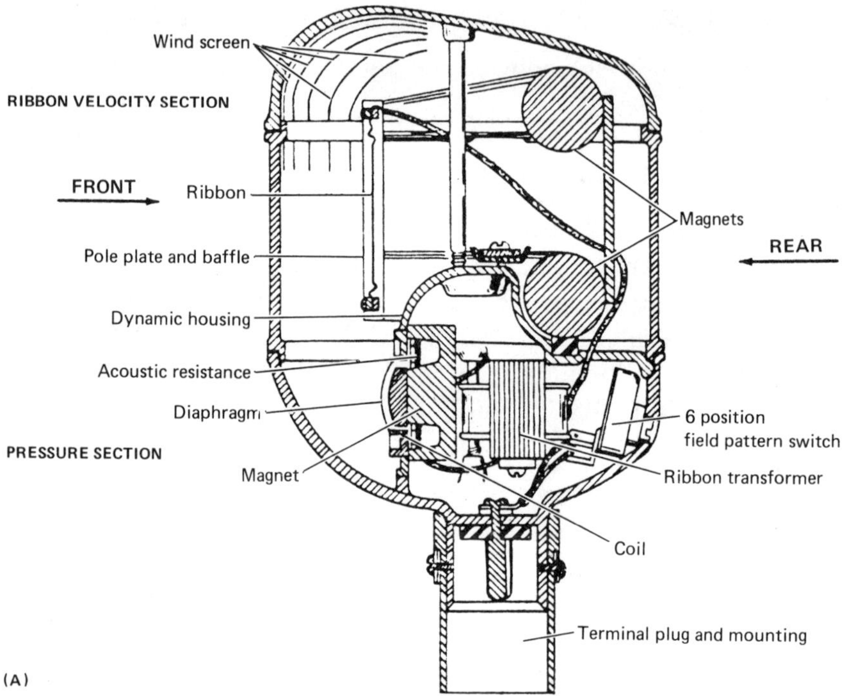

(A)

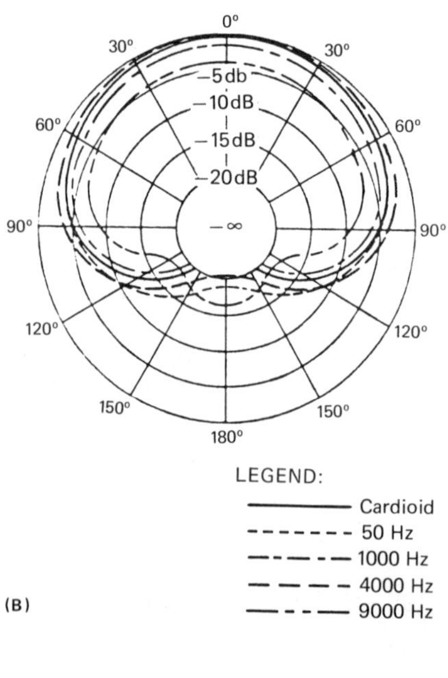

(B)

Figure 2-4. The Western Electric model 639 microphone. **(A)** Detail drawing; **(B)** polar response.

Electrical resonance frequency is determined as

$$f_0 = \frac{\sqrt{1/LC}}{2\pi}$$

Note that the equations are analogous or have the same form.

The amount of resistance or damping in the circuit will determine the shape of the *resonance curve,* as shown in Figure 2-6. Where there is little resistance or damping, the system is said to be stiffness- or capacitance-controlled; the velocity of current will rise as the resonance frequency, f_0, is approached. However, where the resistance or damping is high, the current or velocity will be relatively smooth through the region of resonance. Where there is sufficient resistance or damping in the circuit, this smooth response through the region of resonance is said to be *resistance-controlled:* current or velocity will be proportional only to voltage or pressure (force). Above the resistance-controlled region, the response falls off 6 dB-per-octave with decreasing frequency, and is said to be *mass-* or *inductance-controlled.* In Figure 2-6, note how the response curve flattens out as the circuit damping, or resistance, increases.

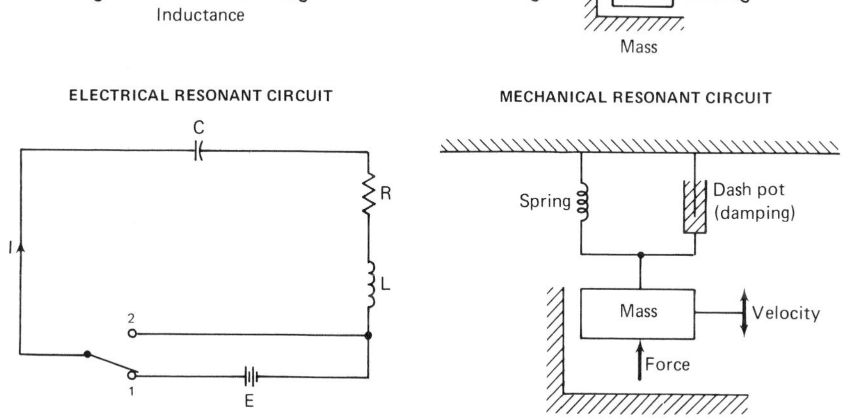

Figure 2-5. Electrical and mechanical equivalent circuits.

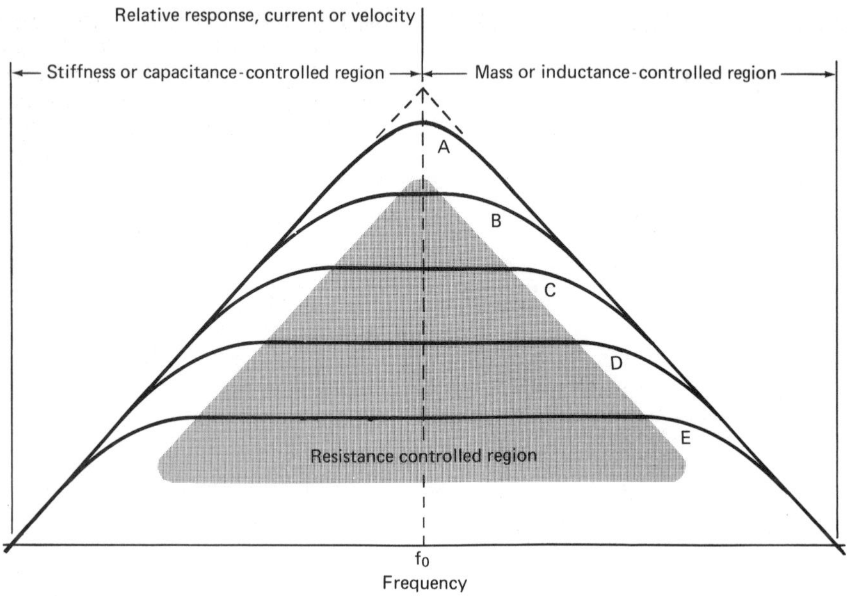

Figure 2-6. The effect of damping on resonance curves. As damping increases, the response flattens out.

THE BRAUNMÜHL-WEBER CONFIGURATION

The Braunmühl-Weber dual element configuration is shown in Figure 2-7. It consists of a stationary back plate, which has perforations through it, and holes on both sides. Diaphragms on either side of the back plate are usually made of metallized mylar plastic. The effect of the perforations depends on the actual movements of the diaphragms, as will be explained shortly. Terminals are attached to both diaphragms, and to the back plate. In effect, the system comprises two identical cardioid elements, in a back-to-back configuration.

We may also describe the Braunmühl-Weber capsule as a combination of pressure and velocity sensing elements as follows. When the elements are connected electrically as shown in Figure 2-8A, the symmetrical bias voltages on the diaphragms produce an output only when they move contrariwise, as shown in Figure 2-8B. Any parallel motion of the diaphragms will produce no output. Thus, in this mode, the diaphragm assembly will respond as a pressure element. The contrary motion of the diaphragm results in no net movement of air through the perforations in the back plate. The enclosed air is alternately compressed and rarefied and the stiffness of the enclosed

(A)

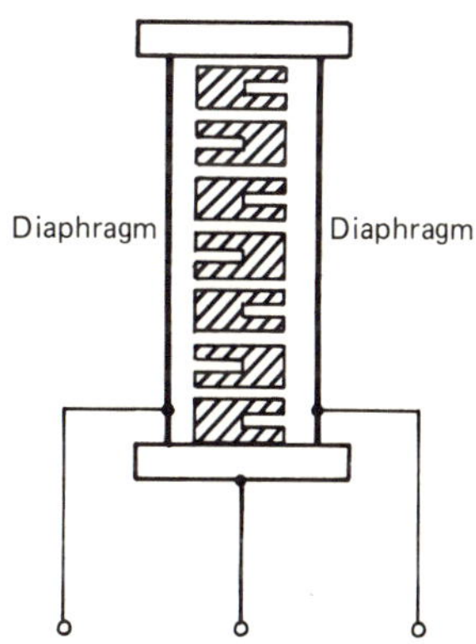

(B)

Figure 2-7. The Neumann U-48 microphone **(A)** utilizes the Braunmühl-Weber principle. *(Gotham Audio photo)* **(B)** Details of the dual-capacitor structure, which may be from 2 to 4 cm in diameter and no more than about 1 cm in thickness.

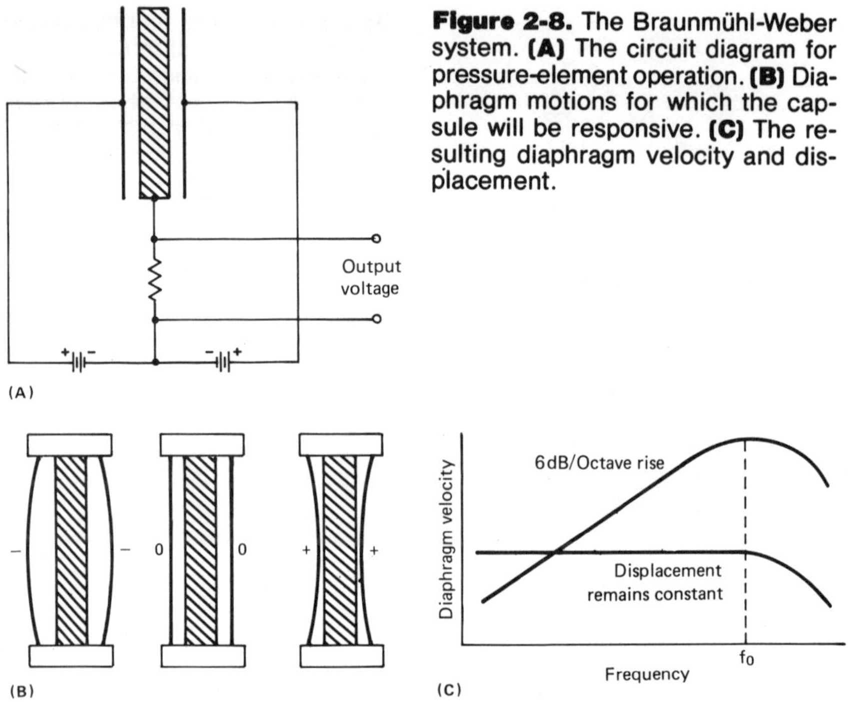

Figure 2-8. The Braunmühl-Weber system. **(A)** The circuit diagram for pressure-element operation. **(B)** Diaphragm motions for which the capsule will be responsive. **(C)** The resulting diaphragm velocity and displacement.

air controls the diaphragm motions. The stiffness control ensures that the diaphragm displacements will be proportional to sound pressure level. As noted before, with each doubling of frequency, the velocity of the diaphragms doubles, and displacement remains constant, as shown in Figure 2-8C.

Now, if we reconnect the bias voltages as shown in Figure 2-9A, the opposite polarities on the two diaphragms will produce an output voltage only they move in parallel motion, as shown in Figure 2-9B. Any contrariwise motion of the diaphragms will produce no output. For parallel motion of the diaphragms, there is no net compression or rarefaction of the enclosed air; rather, the air moves back and forth through the back plate perforations and is thus well-damped, or resistance controlled. Because of the path length difference around the microphone capsule assembly, there will be a net doubling of pressure with each doubling of frequency for any source not located at right angles to the capsule. The velocity rises with frequency, as shown in Figure 2-9C, up to some frequency f_1, where the front-to-back path length is equal to a wavelength. In the better designs, this frequency will be well above the audible range. As noted before, the doubling of velocity translates into a constant displacement with rising frequency,

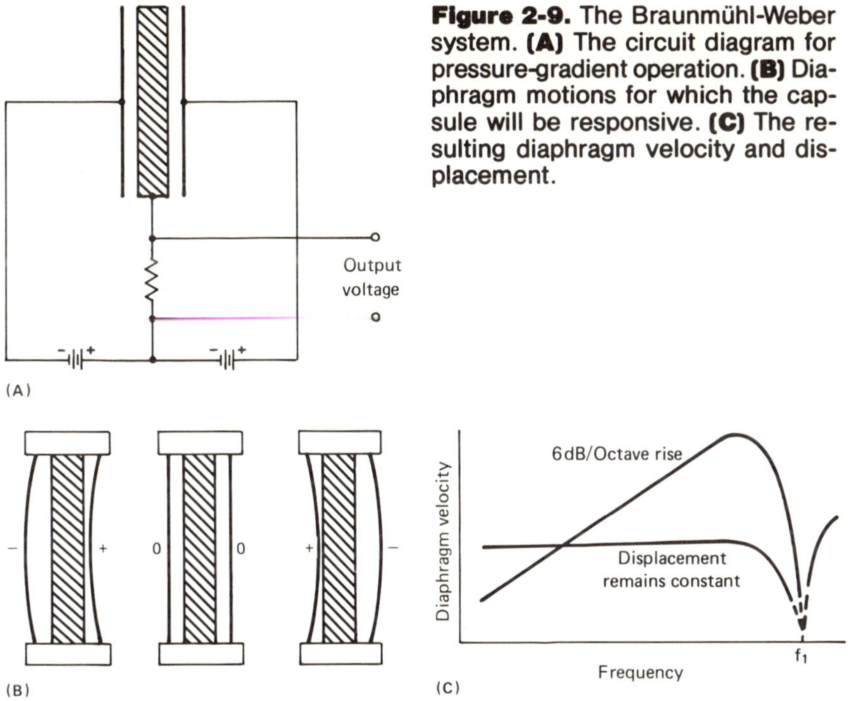

Figure 2-9. The Braunmühl-Weber system. **(A)** The circuit diagram for pressure-gradient operation. **(B)** Diaphragm motions for which the capsule will be responsive. **(C)** The resulting diaphragm velocity and displacement.

and flat output is thus assured. (At this point, we suggest a rereading of the analysis of the ribbon microphone.)

The reader must understand that varying the electrical connections has *no effect* on the actual motion of the diaphragms of the Braunmühl-Weber microphone; it merely alters the output as a function of source direction. The motions of the diaphragms are complex and consist of both parallel and contrary motion. The case is somewhat analogous to a stereophonic phono cartridge. When the two sections are wired together in-phase, there is an electrical output only for lateral motion of the stylus— even though the stylus may be executing complex motion in both vertical and lateral planes. When wired out-of-phase, there is an electrical output only for the vertical component of stylus motion.

With the circuit shown in Figure 2-10, we can combine both pressure and velocity actions of the Braunmühl-Weber assembly and produce the entire family of patterns shown in the figure, which range from figure-8 (Position 1) to omni-directional (Position 5).

The Neumann M-49 microphone illustrates the action explained above. Because of the close spacing of the diaphragms to the back plate and the small overall size of the assembly, the Braunmühl-

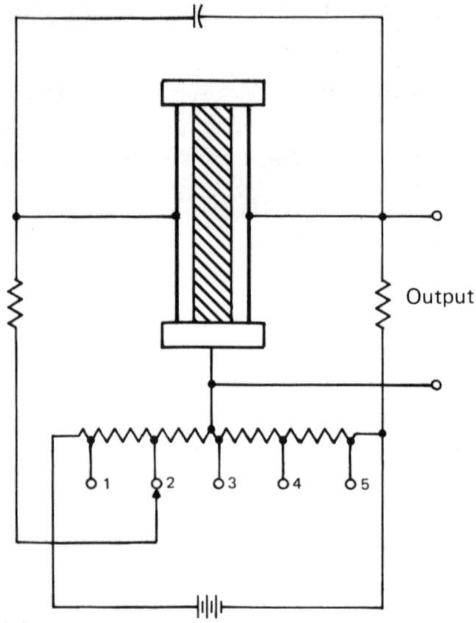

(A)

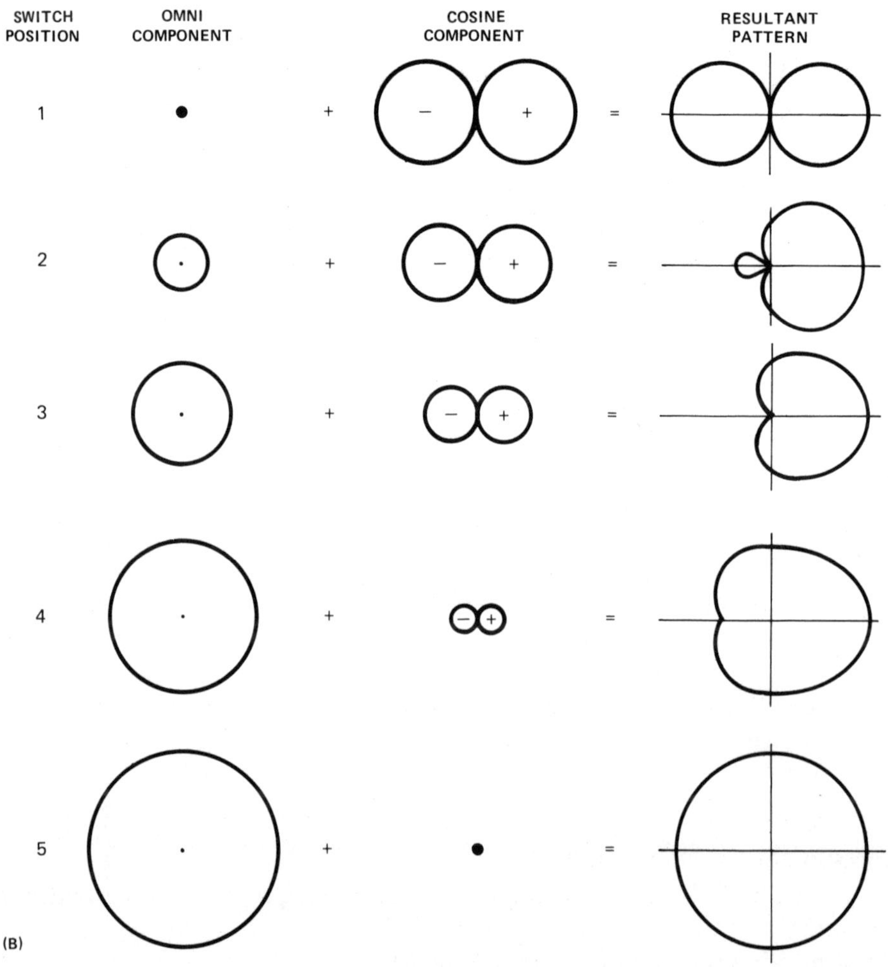

(B)

Figure 2-10. A variable polar-pattern system. **(A)** Circuit diagram; **(B)** resultant polar patterns.

Figure 2-11. The Neumann M-49 microphone *(after Bauch).* **(A)** Capsule detail; **(B)** polar responses; **(C)** frequency response. *(From F. W. O. Bauch, "New High-Grade Condenser Microphone," Jour. of the Audio Engineering Society, 1:3, p. 232 (1953.)*

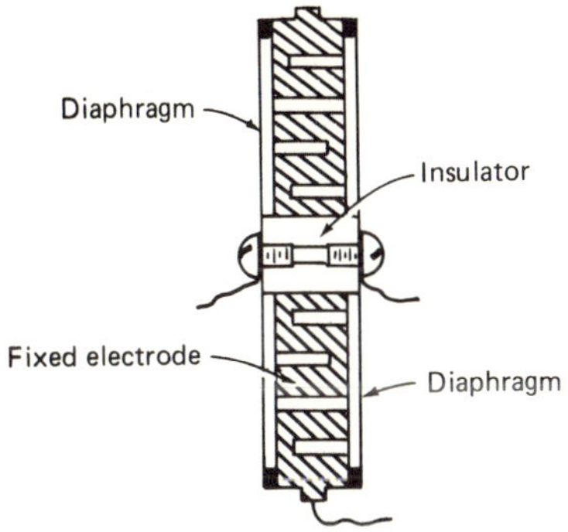

(A)

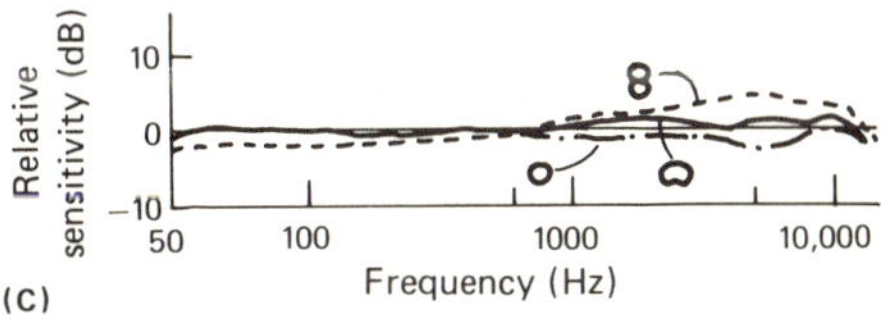

(C)

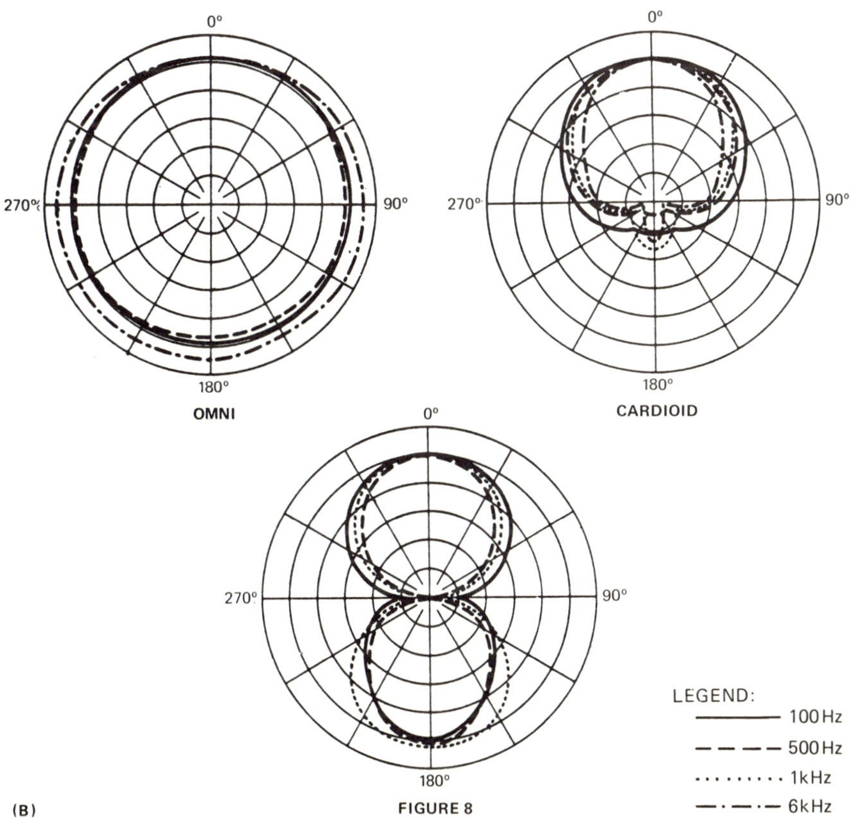

(B)

Weber capacitor microphone maintains excellent directional characteristics, as shown in Figure 2-11.

In addition to the Braunmühl-Weber approach, variable patterns may also be produced by simply combining two back-to-back cardioid microphones, as shown in Figure 2-12.

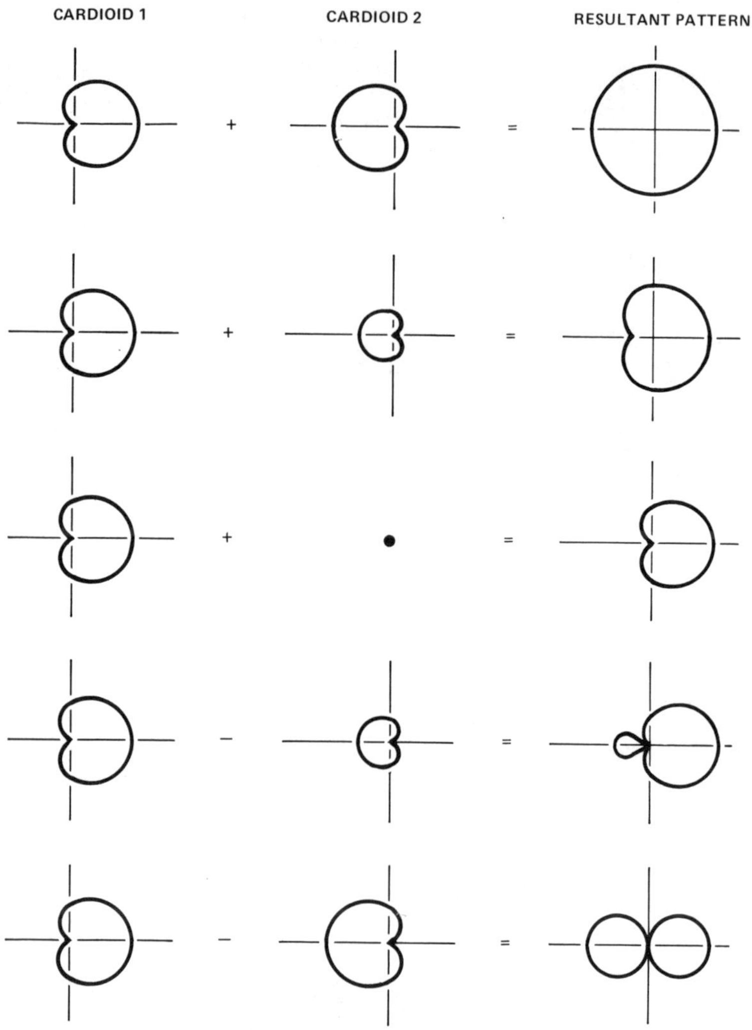

Figure 2-12. A variety of resultant patterns may be obtained by combining the outputs of two back-to-back cardioids.

SINGLE ELEMENT UNIDIRECTIONAL MICROPHONES

Where there is a need to have variable directional response, the dual element approach we have explained is best. However, where only one response pattern is desired, the various single element designs are simplest. In these designs, both pressure and velocity are sensed by a single element through the use of multiple openings on both

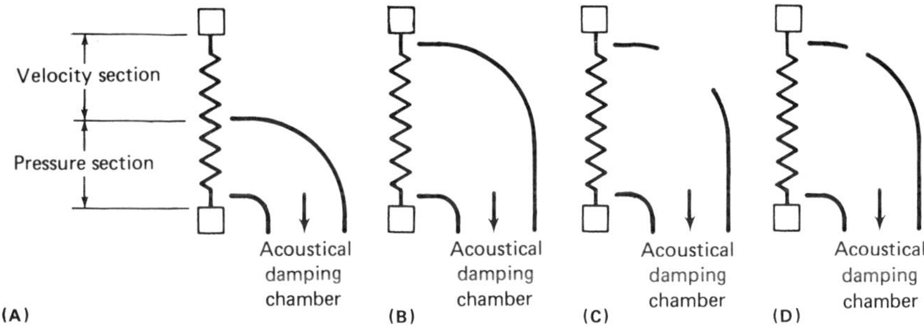

Figure 2-13. A single-ribbon directional microphone (RCA 77 series). **(A)** a fixed uni-directional design. For variable polar response, an adjustable rear-entry port permits **(B)** omni-directional response, **(C)** bi-directional resonse, and **(D)** uni-directional (cardioid) response.

sides of the diaphragm. One of the earliest of these designs was the RCA model 77A microphone. Details of this design are shown in Figure 2-13A. The microphone consists of a single ribbon with a portion exposed in the normal way which provides cosine response. Another portion of the ribbon is shrouded by a cover which leads to an acoustical damping chamber. This portion of the ribbon acts as a pressure device. Thus, the single ribbon element can respond both to velocity and pressure, and a cardioid pattern is produced.

A later version of this microphone operated as shown in Figure 2-13B-D. A shroud with an adjustable opening was placed over the entire ribbon. With the hole closed, the response was omni-directional. When the hole was opened wide, the response was that of a typical ribbon. At some intermediate opening, both pressure and velocity responses were produced in equal amounts, and a cardioid pattern resulted.

Figure 2-14 presents single element dynamic and capacitor designs. The diaphragm, a pressure gradient or velocity device, is arranged so that an acoustical delay at the back of the diaphragm just matches the delay of sound reaching the front of the diaphragm by diffraction around the case. Thus, for sound sources at 180 degrees, there will be no net movement of the diaphragm. As sounds arrive from angles closer to on-axis (0 degrees), there will be progressively less cancellation, and the diaphragm will operate in its normal pressure gradient mode. A precise cardioid pattern results when the path delays, t_1 and t_2, are equal.

In Figure 2-14B and C, we show the "Unidyne" design introduced

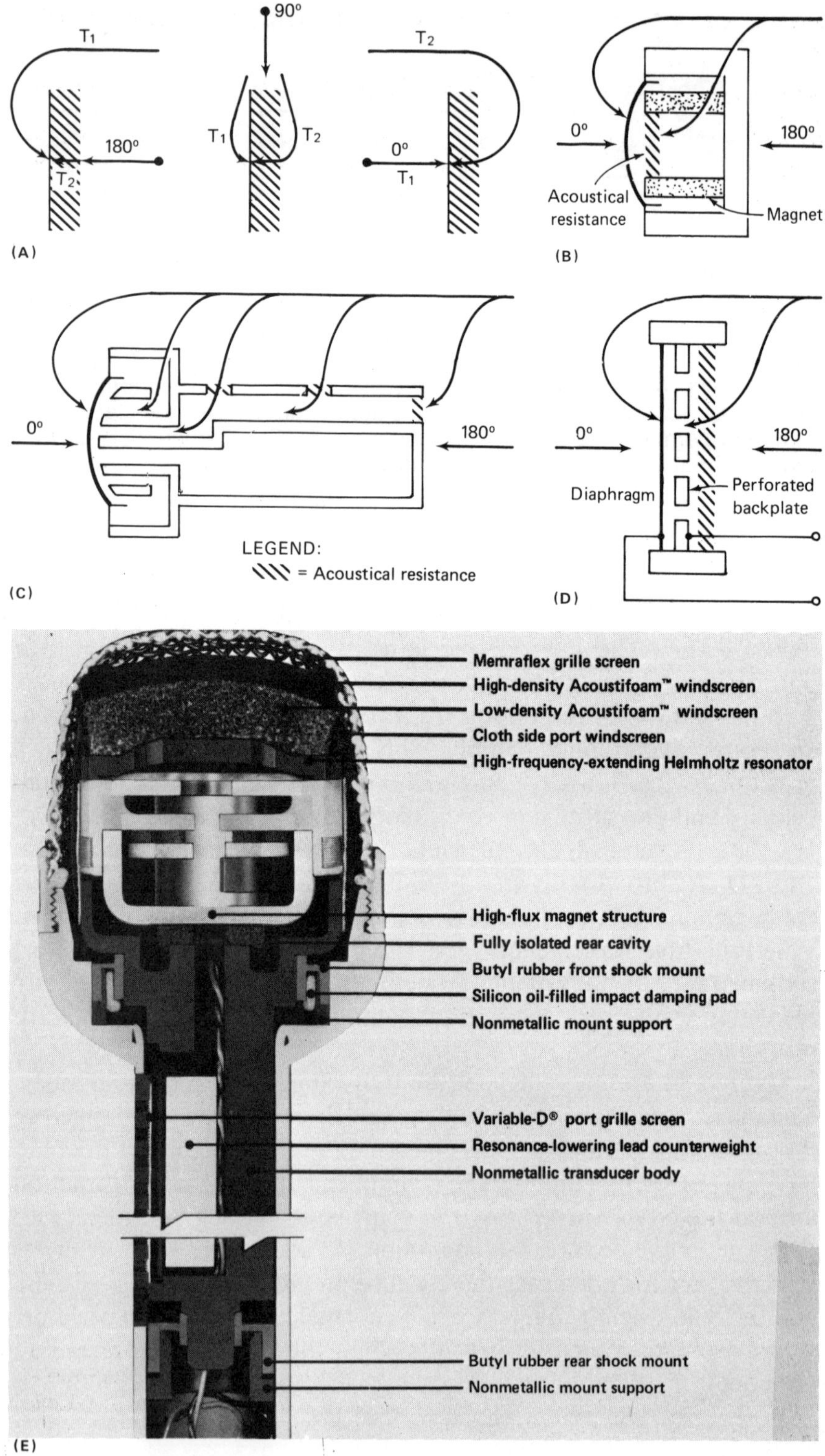
90°
T1
T2
180°
0°
(A)
Acoustical resistance
Magnet
(B)
LEGEND:
= Acoustical resistance
(C)
Diaphragm
Perforated backplate
(D)
Memraflex grille screen
High-density Acoustifoam™ windscreen
Low-density Acoustifoam™ windscreen
Cloth side port windscreen
High-frequency-extending Helmholtz resonator
High-flux magnet structure
Fully isolated rear cavity
Butyl rubber front shock mount
Silicon oil-filled impact damping pad
Nonmetallic mount support
Variable-D® port grille screen
Resonance-lowering lead counterweight
Nonmetallic transducer body
Butyl rubber rear shock mount
Nonmetallic mount support
(E)

Figure 2-14. Cardioid single-element design principles. **(A)** Path length differences for sounds arriving from 180, 90 and 0 degrees. Detail view of **(B)** Shure Unidyne, **(C)** Electro-Voice Variable-D, and **(D)** a typical single-diaphragm capacitor microphone. **(E)** Cutaway view of the Electro-Voice RE 18 Microphone. *(Electro-Voice photo)*

by Shure Brothers, and the Electro-Voice "Variable-D" design. Here, the use of many frequency-selective rear passages minimizes *proximity effect,* a subject to be discussed in detail in Chapter VII.

Finally, a typical single element capacitor design is shown in Figure 2-14D.

The Neumann M-50 microphone is shown in Figure 2-15. This unique design is intended for classical recording, where its combination of omni response at low frequencies and a more-directional response at high frequencies is used to advantage. It consists of a single pressure capacitor element mounted on a plastic sphere. The transition to directional response above 2 kHz gives more delineation to high frequencies, as does the rise in output level above 3 kHz. At low frequencies, sound bends around the sphere and pressure response results. At high frequencies, the sphere acts as a baffle, and the capacitor element responds more to signals originating along its principal axis. The pattern of the M-50 allows it to be placed some

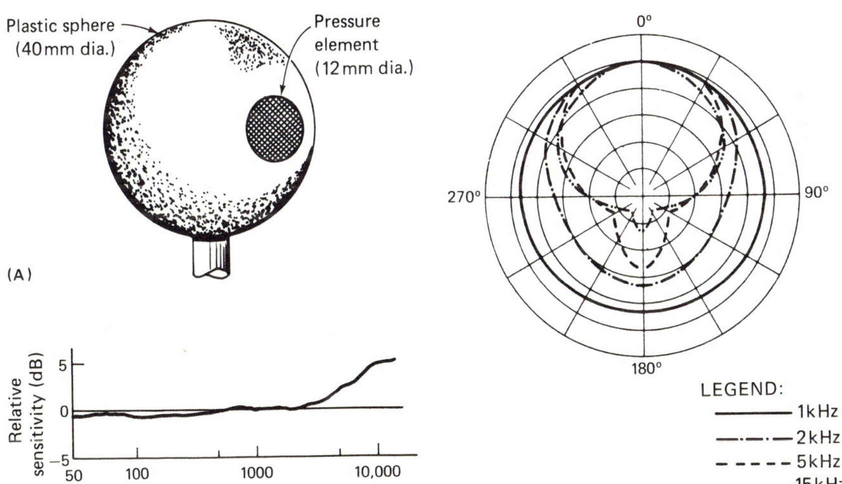

Figure 2-15. The Neumann M-50 microphone *(After Bauch).* **(A)** The microphone element; **(B)** polar response; **(C)** on-axis frequency response. *(From F. W. O. Bauch, "New High-Grade Condenser Microphone," Jour. of the Audio Engineering Society, 1:3, p. 232 (1953.)*

distance from the orchestra while preserving a degree of intimacy due to the rising HF response and in giving additional musical detail through the use of a cardioid pattern at high frequencies.

THREE-DIMENSIONAL REPRESENTATION OF POLAR RESPONSE

Polar response of all microphones exists in three dimensions, while the convenient polar plot is only two-dimensional. One can imagine the polar plot as rotated about the major axis in order to produce a three-dimensional representation. Examples of this are shown in Figure 2-16.

RANDOM ENERGY EFFICIENCY (REE) OF MICROPHONES

The engineer soon learns that directional microphones do not completely reject sounds originating along their null angles. What directional microphones effectively do is control the amount of *random* signal level relative to the *desired* signal originating along the principal axis. A convenient measure of the ability of microphones to do this is its REE; the more directional the microphone, the lower its REE.

An omni-directional microphone, since it responds equally well in all directions, has a REE of 1; a cardioid microphone, set at the same level for on-axis pickup, will only pick up one-third as much random energy. Thus, its REE is 0.33. We can demonstrate this as shown in Figure 2-17. If both omni and cardioid microphones are set for the

(A) (B)

Figure 2-16. Three-dimensional representations of polar patterns for an **(A)** omni-directional, and **(B)** uni-directional microphone. *(Data courtesy Audio-Technica.)*

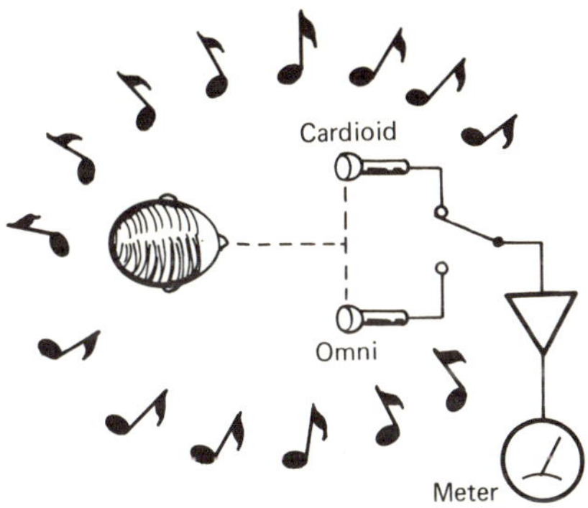

Figure 2-17. Random Energy Efficiency (REE). When both microphones are equidistant from a speaker, and are adjusted for equal level, the ambient noise level picked up by the cardioid will be one-third that of the omni (−4.8 dB).

same response level to a speaker located close by, the reverberant noise level detected when the speaker is silent will be one-third or 4.8 dB lower, through the cardioid microphone as compared to the omni. A *hyper-cardioid* pattern minimizes the REE, while a super-cardioid pattern has the greatest ratio of frontal pickup (±90 degrees) to reverberant pickup.

The experiment shown in Figure 2-17 works best in a highly reverberant space. In a typical recording studio, the reverberant field would be quite low, and the REE would simply manifest itself as a specific amount of off-axis attenuation, as observed from the microphone's polar pattern. The REE concept is most useful in sound reinforcement applications in reverberant rooms where specific feedback paths around the electroacoustical system cannot be accurately predicted. In these cases, the REE concept can give the system design engineer an indication of how effective a given pattern will be in maintaining some feedback stability margin. This topic will be covered in detail in Chapter XI.

The microphones we have studied so far in this chapter have been essentially *first-order* devices, that is, they have been constructed basically on the simple mathematical models of omni ($\rho = 1$) and figure-8 ($\rho = \cos\theta$) curves.

Figure 2-18 presents a summary of first-order microphone data. Note the heading called *distance factor;* this establishes the distance, with respect to the omni pattern as unity, at which the direcitonal pattern may be used for the same REE. For example, a cardioid pattern operated at a distance of 1.7D will, after on-axis gains have been equalized, result in the same random energy level as an omni pattern operated at a distance, D. Note that the hyper- and the super-cardioid allow up to *two times* the operating distance of an omni for a fixed ratio of on-axis to random energy response.

CHARACTERISTIC	OMNI-DIRECTIONAL	CARDIOID	SUPER-CARDIOID	HYPER-CARDIOID	BIDIRECTIONAL
Polar response pattern	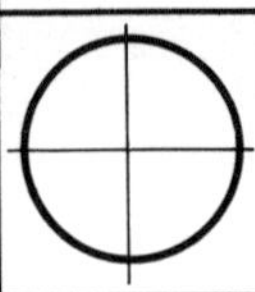	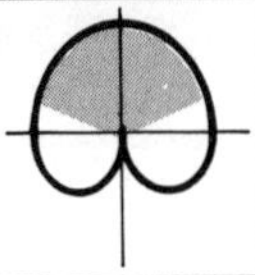	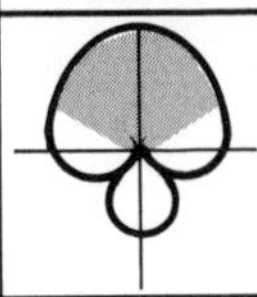	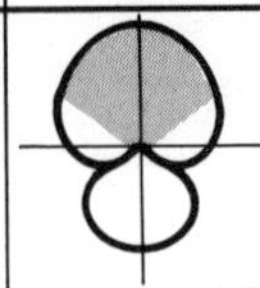	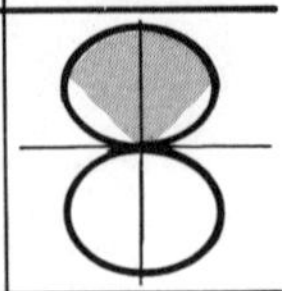
Polar equation	1	$.5 + .5 \cos\theta$	$.375 + .625 \cos\theta$	$.25 + .75 \cos\theta$	$\cos\theta$
Pickup ARC 3 dB down (1)	—	131°	115°	105°	90°
Pickup ARC 6dB down	—	180°	156°	141°	120°
Relative output at 90° dB	0	−6	−8.6	−12	−∞
Relative output at 180° dB	0	−∞	−11.7	−6	0
Angle at which output = 0	—	180°	126°	110°	90°
Random energy efficiency (REE)	1 0dB	.333 −4.8dB	.268 −5.7dB (2)	.250 −6.0dB (3)	.333 −4.8dB
Distance factor (DF)	1	1.7	1.9	2	1.7

NOTE:
1 = Drawn shaded on polar pattern
2 = Maximum front-to-total random energy efficiency for a first order cardioid
3 = Minimum random energy efficiency for a first order cardioid

Figure 2-18. A summary of first-order microphone data.

HIGHER-ORDER DESIGNS

For special applications, "shot-gun" or parabolic microphones may be used. These are designs which are wavelength dependent in their operation and provide high directionality at higher frequencies. A shot-gun design is shown in Figure 2-19. Sounds arriving along the major axis enter the microphone relatively unimpeded, while those entering at angles progressively off axis will be subject to more cancellation. The *larger* the structure, the lower will be the frequency at which the design exhibits its desired high directivity. Typical large designs, while cumbersome to use, work well down to 200 Hz and provide REEs on the order of 0.1.

A large parabolic dish with an omni microphone located at the focus, as shown in Figure 2-20, will demonstrate excellent directional

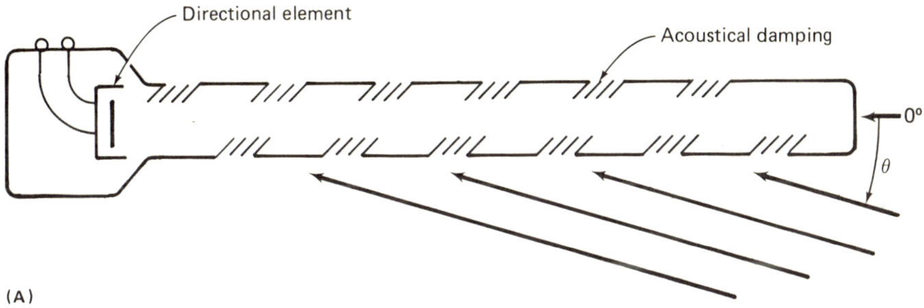

(A)

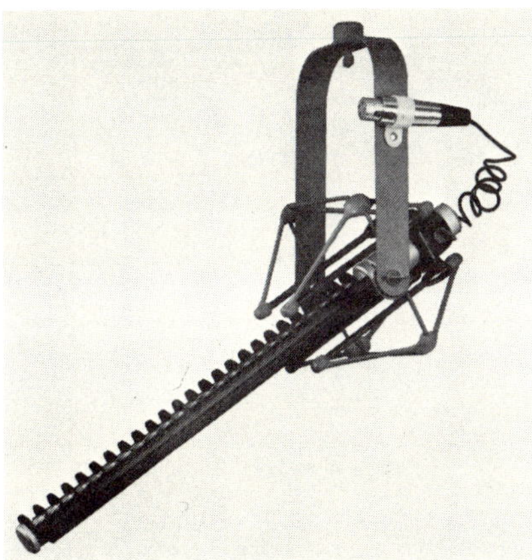

(B)

Figure 2-19. The shot gun or line microphone. **(A)** Sounds entering at zero degrees reach the microphone element with less attenuation than sounds incident at off-axis angles. **(B)** The Electro-Voice CL42S microphone *(Electro-Voice photo)*

characteristics. Again, the larger the dish, the lower in frequency the device will work.

The parabolic dish is very impractical for general use but is excellent for certain kinds of surveillance work or recording in nature.

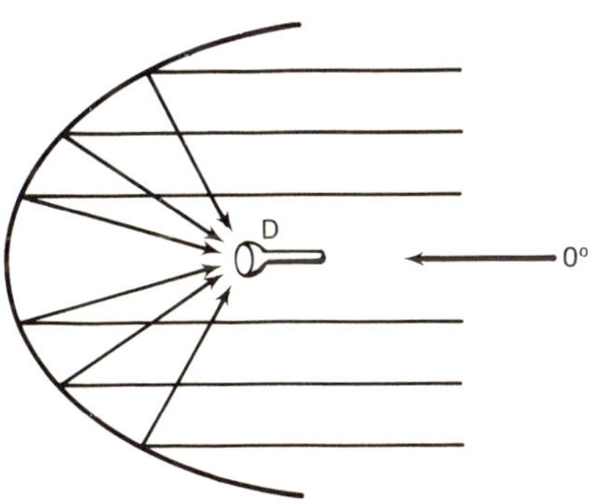

Figure 2-20. Principle of the parabolic microphone. The parabolic disc gathers sound waves at middle and high frequencies and focuses them at the microphone element.

CONCLUSIONS

In this chapter, we have discussed the various ways directional response in microphones can be produced. The reader should take note of the roles played by pressure response and velocity response in microphones, whether they are embodied in a single element or a pair of elements. When a microphone exhibits any degree of unidirectionality, it is always as a result of interaction of pressure and velocity response.

CHAPTER III

Microphone Sensitivity Ratings

Microphone sensitivity ratings state the open-circuit output voltage, or output power available, when the microphone is placed in a sound field of specified level. Over the years, a number of rating methods have been used, and a survey of current manufacturers' data sheets shows that there is no single method accepted by the industry.

OPEN-CIRCUIT VOLTAGE RATING

This rating method has become popular in recent years, as more microphone input circuits are designed to present a high-impedance bridging load to the microphone. Usually, the reference sound field is specified as either 74 dB-SPL (1 microbar, or 1 dyne/ cm^2), or 94 dB-SPL (10 microbars, or 1 pascal).

If the microphone's response is not substantially flat, the sensitivity rating will be stated at some frequency, usually 1 kHz. A typical specification might read

Open Circuit Voltage = −80 dB (0 dB = 1 volt/ microbar).

This means that when the microphone is placed in a 74 dB-SPL sound field, the resulting output voltage will be 80 dB *below* 1 volt. Details of this are shown in Figure 3-1. Since the microphone is unloaded, there will be no voltage drop across the effective source

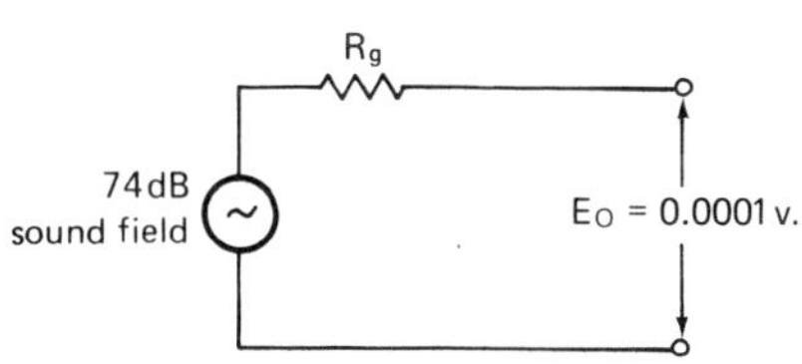

Figure 3-1. The equivalent circuit of an unloaded microphone.

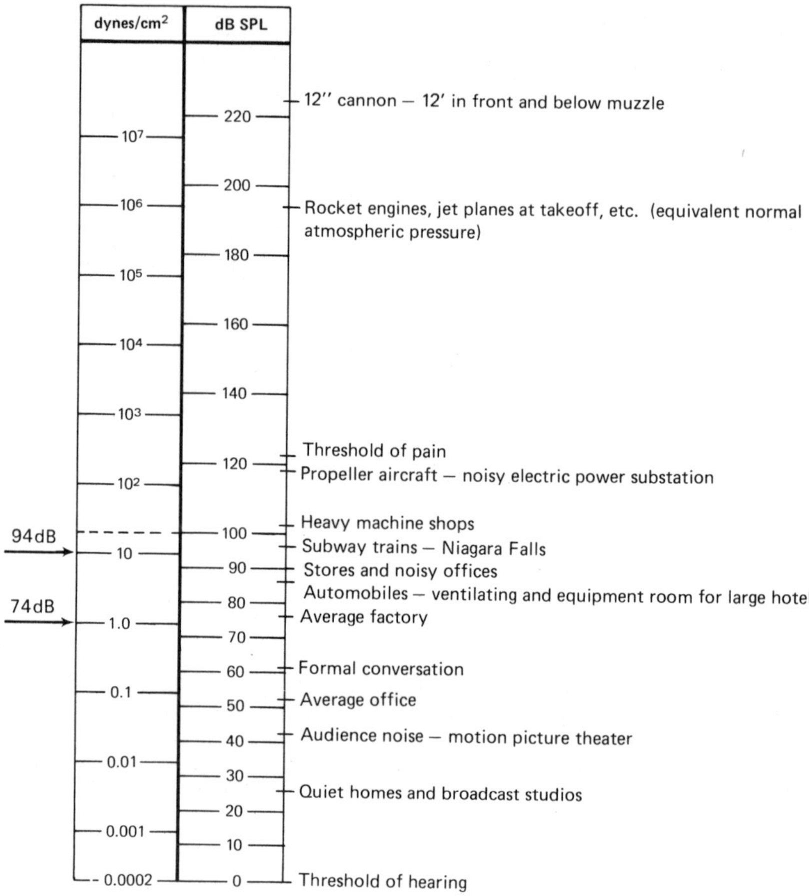

Figure 3-2. Typical sound pressure levels.

resistance of the microphone, R_g, and full output will appear at the terminals.

We can compute the output voltage as follows:

$$E_0 = \text{antilog}_{10}\left(\frac{\text{level in dB}}{20}\right)$$

$$= 10^{-80/20} = 10^{-4} = 0.0001 \text{ volts}$$

As another example:

Open Circuit Voltage $= -73$ dB (0 dB = 1 volt/microbar).
$E_0 = \text{antilog}_{10}(-73/20) = 0.224 \times 10^{-3} = 0.000224$ volts.

Figure 3-2 shows some typical sound sources and their corresponding sound pressure levels. Note that the standard reference levels of 74 and 94 dB-SPL represent levels encountered in normal use. When comparing specifications, note carefully which reference sound pressure level is used. A microphone rated at −60 dB (1 volt/microbar) will be equivalent to −80 dB (1 volt/10 microbars).

Figure 3-3 gives voltage values corresponding to levels in dB below 1 volt. The term dBV is often used to indicate these levels.

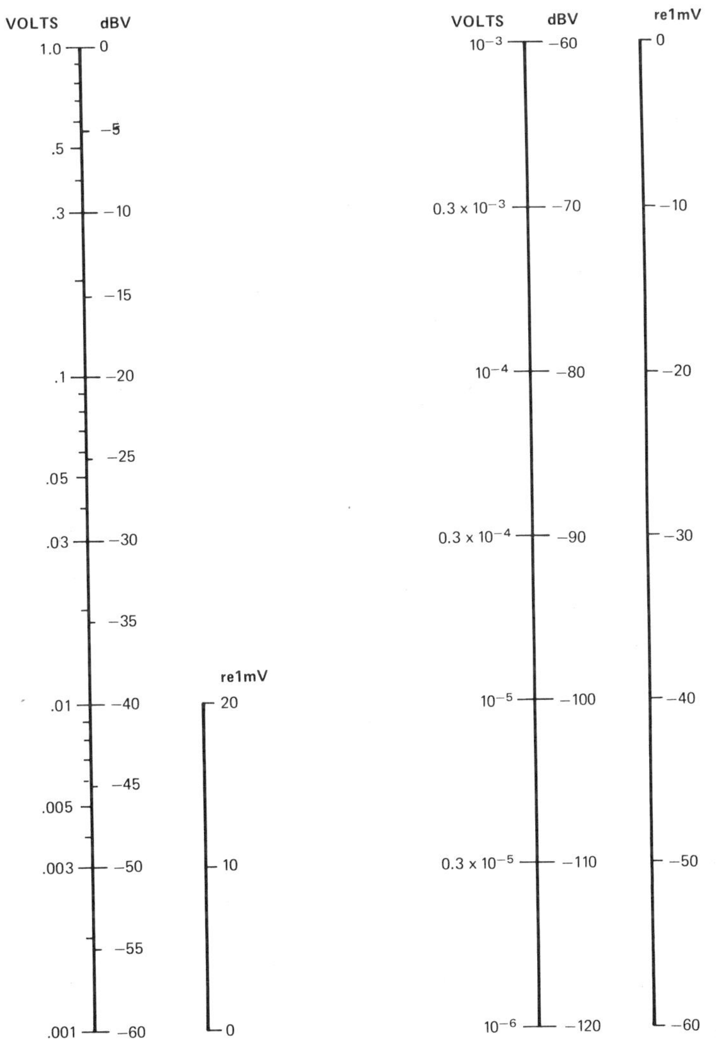

Figure 3-3. A comparison of decibel levels and voltage (0 dBV = 1 volt).

MICROPHONE POWER RATINGS

The amount of power generated by a microphone is extremely small, and microphone power ratings are a holdover from an earlier day when it was customary to match impedances of devices in an audio chain for maximum power transfer. Let us refer to Figure 3-4A, where the microphone is unloaded and the full output voltage appears at the microphone's terminals. In Figure 3-4B, the microphone has been loaded with a resistance, R_L, equal to the microphone's own internal resistance, R_g. In this case, the output voltage drops to *one-half,* since the voltage drop is the same across both resistances. The output power is

$$\frac{(E_0/2)^2}{R_L} = \frac{E_0^2}{4R_L}$$

In low-level transmission, power levels are almost always expressed in dB relative to one milliwatt (mW). The designation, dBm, is used to indicate that level.

Thus, Level in dBm = 10 log(P/.001), where

P = any power.

A reference sound pressure level of 94 dB-SPL (10 dynes/cm² or 1 pascal) is invariably used when a microphone power output specification is given. A typical specification might read

Impedance = 200 ohms,
Power Output = −40 dBm (10 dynes/cm²).

First, let us determine the actual power:

$$P = \text{Antilog}_{10}(-40/10) = 10^{-4} = 0.0001 \text{ mW}$$
$$= 0.1 \times 10^{-6} \text{ watts}$$

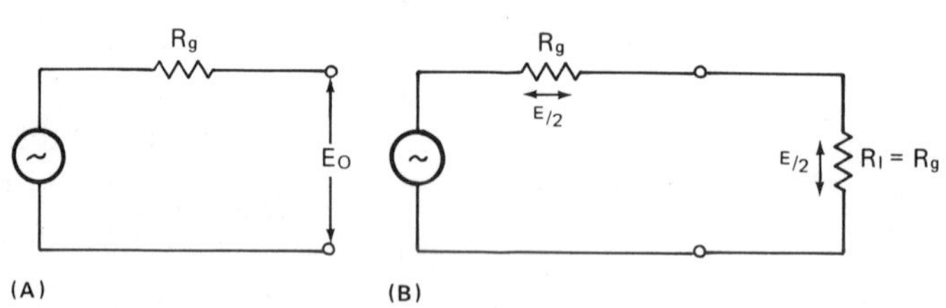

Figure 3-4. The equivalent circuit **(A)** of an unloaded microphone; **(B)** of a microphone loaded with a resistance, R_1.

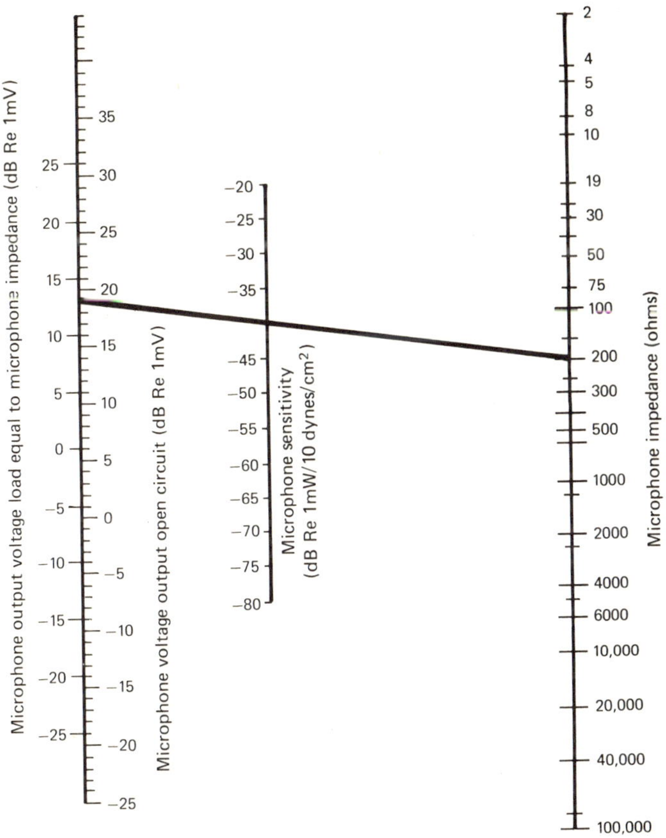

Figure 3-5. Nomograph of microphone impedance, output power, and voltage level.

Finding the voltage across 200 ohms:

$$P = \frac{E^2}{200} = 0.1 \times 10^{-6}$$

$$E^2 = 20 \times 10^{-6}$$
$$E = 4.47 \times 10^{-3} \text{ volts}$$

In this expression, E is the voltage across a load of 200 ohms. The unloaded, or open-circuit, voltage would be twice this, or 8.94×10^{-3} volts. While these calculations are straightforward enough, it is far simpler to use the nomograph shown in Figure 3-5.

Another power output rating method was devised by the Electronic Industries Association (EIA). The term G_m is used to indicate the power output level in dBm for a sound pressure level of 0 dB-SPL. The microphone impedance value in the EIA rating is a nominal value chosen from the following series: 38 ohms, 150 ohms, 600 ohms, 2400 ohms, etc., where each higher value is four times that of the previous value. The range of each nominal value is from one-half to twice the nominal value. Thus, in the preceding example, the actual microphone impedance of 200 ohms would fall into the nominal 150 ohm range (75-to-300 ohms).

We will now calculate the G_m rating for our previous example. In that example, a sound pressure level of 94 dB produced an open-circuit voltage of 8.94×10^{-3} volts. For a sound pressure level of 0 dB, the output voltage will be 94 dB lower:

$$E = 8.94 \times 10^{-3} \times 10^{-94/20} = 1.78 \times 10^{-7} \text{ volts}$$

When loaded, the voltage drops by one-half (0.89×10^{-7} volts). The power is then

$$P = \frac{E^2}{150} = 5.28 \times 10^{-14} = -132.77 \text{ dBm}$$

If we had simply ignored the change from 200 ohms to the nominal 150 ohm rating and subtracted 94 dB from the power rating (re 94 dB-SPL), we would have $-40 - 90 = -134$ dB.

As a general rule, we can simply add 94 to a G_m rating to find the effective rating referred to 1 mW/10 dynes/cm^2.

OUTPUT TRANSFORMERS

Some microphones with output transformers have provisions for strapping those transformers two ways, as shown in Figure 3-6. While the voltage ratio is 2:1, the impedance ratio is 4:1. Thus, a 200-ohm microphone with its output windings in series can be converted to a 50-ohm microphone by strapping the windings in parallel. The 6-dB difference in output voltage level offers the user a degree of flexibility in matching various input conditions.

While the open-circuit voltage rating for a given microphone will vary by 6 dB, depending on the strapping of the transformer, the output power rating of the microphone will be the same, regardless of the transformer strapping.

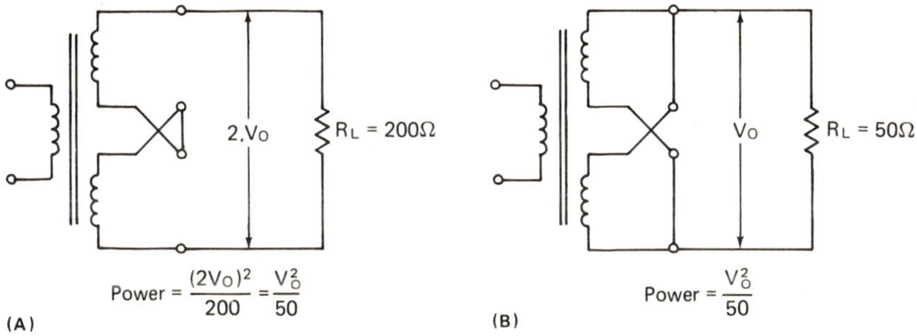

Figure 3-6. Output transformer strapping. The microphone may be wired for **(A)** 200-ohm (series) or **(B)** 50-ohm (parallel) operation.

MICROPHONE NOISE RATINGS

As a general rule, microphones are not offenders in terms of noise; problems in the division of gains and losses are usually the culprits when noise is encountered. However, microphones do have fundamental noise limits, and care must be taken to ensure that the wrong microphone is not specified for a particular application. The preamplifiers which are part of all capacitor microphones deserve special attention at the design stage to ensure that noise is minimized. Ignoring for the present time the problems of shielding against hum and radio-frequency interference (RFI), there is a fundamental noise limit for a microphone due to thermal agitation of molecules in the structure of its transducing parts. The open-circuit value of this noise is

$$E = \sqrt{4kRT\Delta f}$$

where

k = Boltzmann's constant (1.38×10^{-23} joules/°Kelvin),
R = The microphone's source resistance,
T = Temperature (in degrees Kelvin), and
Δf = Frequency bandwidth.

A microphone with a source resistance of 50 ohms, operating at a temperature of 300 degrees Kelvin, and with a bandwidth of 20 kHz, will have a noise voltage of 12.8×10^{-8} volts, or 0.13 microvolts.

Expressing this as a noise power across a matching load resistance (itself a noise generator at these low levels), we have an effective noise power level of -125 dBm.

The calculation is made as follows: when the 50-ohm microphone is loaded with a matching 50-ohm resistor, the combined resistance across the signal output is 25 ohms (the parallel combination of two 50-ohm resistances). Thus

$E = \sqrt{4kRT\Delta f}$, and therefore,
$P = E_0^2/R = 4kT\Delta f.$

The level (referred to 1 milliwatt) is
$N_{dB} = 10 \log(4kT\Delta f/0.001) = -125$ dBm.

Let us assume that a particular dynamic microphone has a power output rating of −60 dBm (10 dynes/cm², 94 dB-SPL). Then, the effective "noise floor" of this microphone will be 125 − 60 = 65 dB below 94 dB-SPL, or 29 dB-SPL. Such a microphone would be useful in fairly quiet environments down to the 30-to-35 dB-SPL range. Referring to Figure 3-2, we note that this is in the region of the typical noise level of a broadcast or recording studio.

Often, the residual noise of a microphone is measured via an "A-weighting" curve, as shown in Figure 3-7. These readings are referred to as dBA levels. This curve takes into account the relative insensitivity of the ear to low-frequency sounds at extremely low levels. Accordingly, a microphone with an inherent noise level 65 dB below 94 dB-SPL (29 dB-SPL) would read lower still in noise, due to the weighting network. Typically, we would expect to see dBA ratings in the 15-to-20 dBA range.

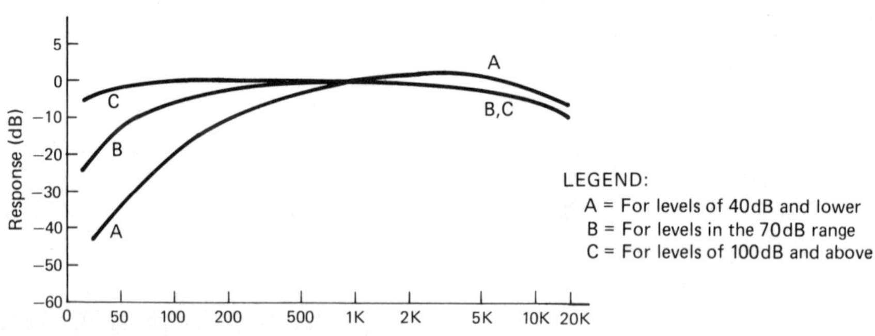

Figure 3-7. Weighting curves for sound level measurements of **(A)** 40 dB and lower, **(B)** in the 70 dB range, and **(C)** 100 dB and higher.

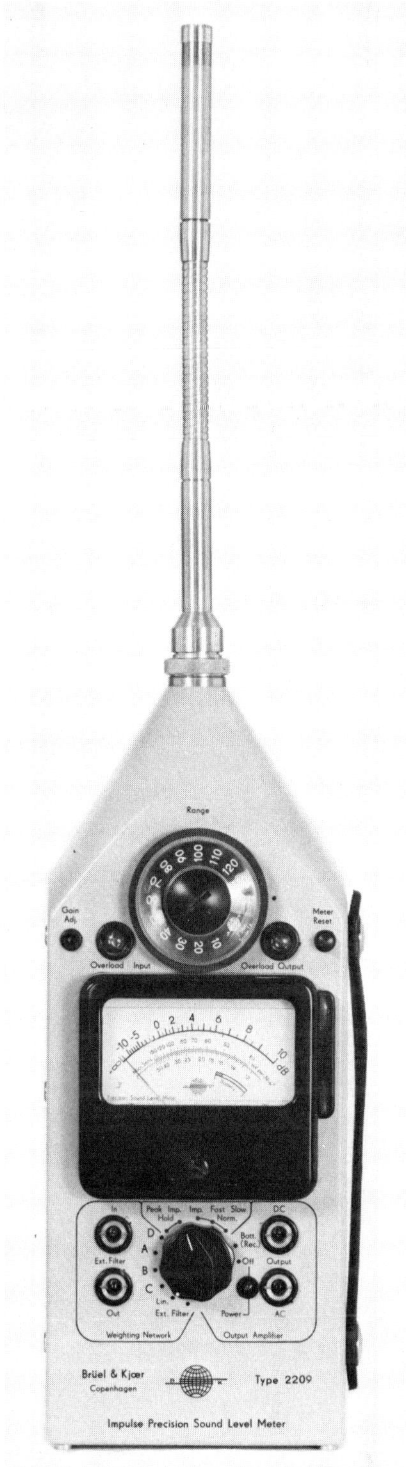

Figure 3-8. A typical high quality sound level meter *(Bruel & Kjaer photo).*

Most microphones can respond in linear fashion well into sound pressure levels of 120-to-130 dB. As a result, the typical professional microphone can accommodate a total dynamic range approaching 100-to-105 dB, better than even the best recording devices used today.

SUMMARY

In comparing open-circuit voltage ratings of microphones, note carefully the reference sound pressure level. Levels of 74 dB-SPL (1 dyne/cm^2) and 94 dB-SPL (1 pascal) are used about equally. Comparisons can be made between ratings, if the 20 dB difference in level is taken into account.

Power ratings will invariably be given in dBm, referred to 94 dB-SPL (0 dB = 1 mW/10 microbars). A conversion may be made from the rarely-encountered EIA G_m rating effectively by adding 94 dB to that rating.

If the microphone has been chosen for the job at hand, noise due to microphone source impedance or to the preamplifier of a well-designed capacitor microphone should not be a problem. Usually, noises due to induction (hum) and RFI are more obtrusive.

CHAPTER IV

Remote Powering of Capacitor Microphones

INTRODUCTION

All types of capacitor microphones require power at the microphone capsule itself. Standard capacitor microphones require polarizing voltage as well as power for the integral preamplifier. While electret and RF types do not require external polarization, they do require power for preamplifiers and associated electronics. Older tube-type systems required a filament voltage supply as well.

For years, recording engineers had to put up with the nuisance of external power supplies for capacitor microphones. The usual rule was one supply per microphone, and the cable connecting the supply to the microphone housing contained multiple conductors within a single shield. A typical circuit diagram for an older power supply is shown in Figure 4-1.

CIRCUITS FOR REMOTE POWERING

In remote powering, the modulation leads are used for supplying power to the microphone. The terms *multiplex powering* and *phantom powering* also describe this. An early circuit used only the signal leads

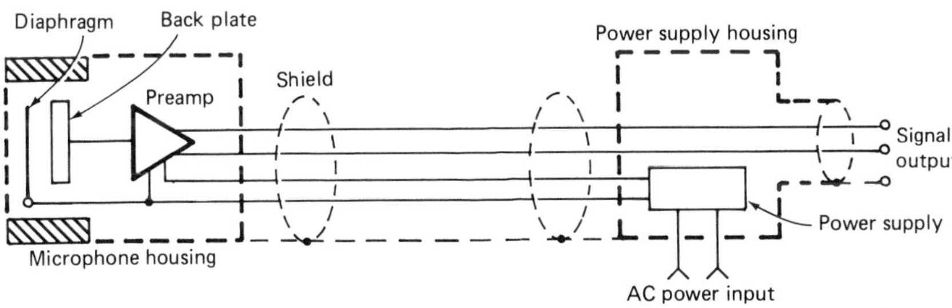

Figure 4-1. Conventional powering of capacitor microphones.

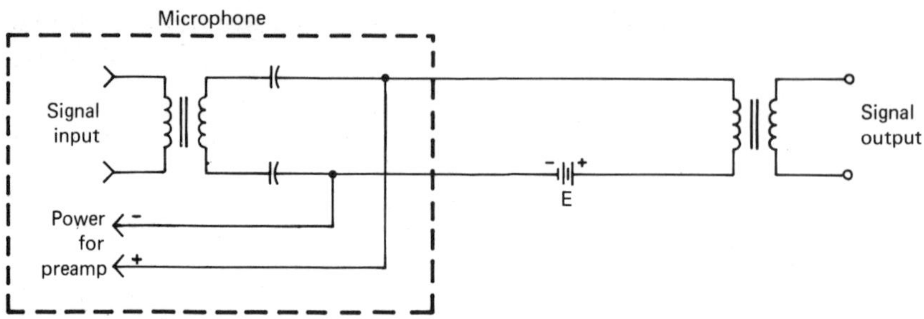

Figure 4-2. An early method for remote powering.

for power transfer, as shown in Figure 4-2. The advantage of this method is that the shield does not carry current; a disadvantage is that the DC supply must be extremely well-filtered, since both signal and DC voltages exist in the same leads. The method has the further disadvantage of requiring careful decoupling in the power supply if more than one microphone is used and if crosstalk is to be minimized.

If a dynamic or ribbon microphone is connected to a system equipped for this kind of remote powering, the power supply will have to be disconnected in order to prevent the flow of current through the voice coils or transformers which are part of these microphones. The presence of DC under these circumstances will result in distortion and possibly, microphone damage.

In recent years, a method of remote powering has become standard in which both signal leads are at the same positive voltage, while the shield is used as the DC return path. In Figure 4-3A, the positive terminal of a battery or DC power supply is fed to the center tap at the console input transformer. In Figure 4-3B, an artificial, or virtual center tap is created through the use of two carefully-matched resistors.

An advantage of this remote powering method is that dynamic and ribbon microphones may be used with no degradation in performance; since both signal leads are at the same voltage and there is no return path, there will be no direct current flow through the signal leads.

For multiple microphone use, the circuit of Figure 4-3B can be implemented in a console, as shown in Figure 4-4. As in Figure 4-3B, the values of R must be maintained within a tolerance of one percent.

While individual capacitor microphones may vary in their nominal voltage requirements, most are fairly tolerant of minor deviations from nominal. If the current drain for a given microphone is known, the voltage drop through the resistors of Figure 4-3B can be calculated simply as IR.

For a value of R = 6800 ohms and a current drain of 0.5 milliamperes, the voltage drop across the series-parallel combination of the four resistors will be:

$$V = 6800\ (0.5 \times 10^{-3}) = 3.4 \text{ volts}$$

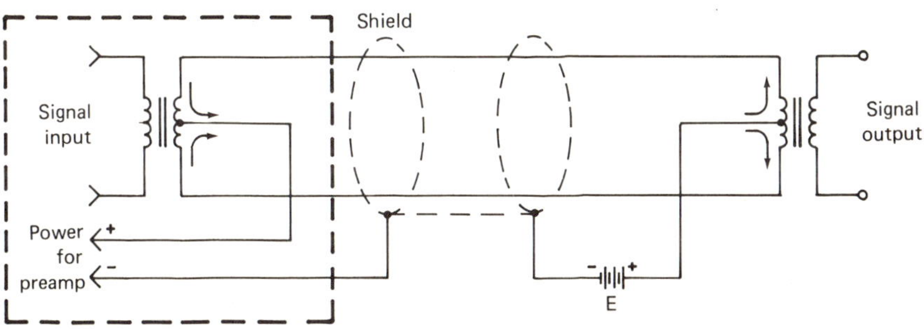

(A)

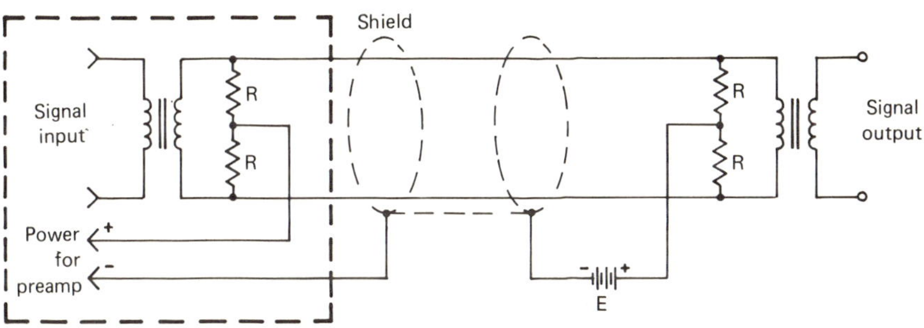

(B)

(C)

Figure 4-3. Remote powering, with both signal leads positive. The power supply's positive terminal may be connected to the transformer center tap **(A),** or to a resistor network. **(B).** A battery supply **(C)** may also be used for remote powering applications *(Electro-Voice photo).*

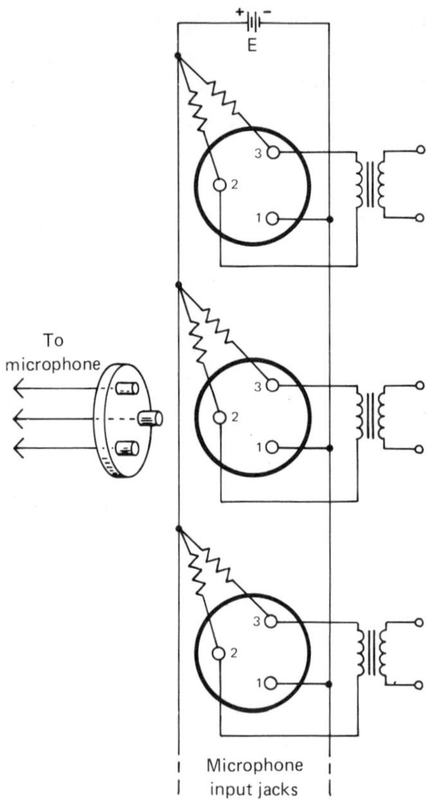

Figure 4-4. A multiple-input remote powering system.

These values of resistance and current drain are typical, and the resulting voltage drop may be considered insignificant. We must further consider the effect of the series combination of the resistors operating in parallel with the primary of the console input transformer. Typical resistance values are high enough relative to the normal impedances seen by the audio signal, and the effect of the added resistance across the input transformer is negligible.

Supply voltages normally used in remote powering are 12 volts to 48 volts. The lower voltage is sufficient for powering the FET (field effect transistor) amplifiers associated with electret microphones. Through the use of DC-to-DC converters, lower voltage levels can be converted upward to higher values for those microphones which may require them. The 48-volt value is sufficient for direct polarization of current capacitor microphone designs. Through the use of internal voltage regulation, an electret microphone may work over the entire range from 12 to 48 volts.

GENERAL RECOMMENDATIONS

Remote powering offers great convenience, and no studio system should be contemplated without it. It requires that the studio manager survey carefully the range of microphones to be used so that they will all be compatible with the basic power supply. Of course, older tube-type capacitor microphones will still require their original power supplies, because of their relatively high polarizing voltages as well as filament voltage requirements.

Since the microphone cable shield is a part of the powering system, absolute continuity in the shield is essential. Thus, maintenance of cable and contact integrity is important. Variations in shield resistance, a common occurence in much-used cables, will not necessarily affect the performance of dynamic or ribbon microphones. However, these variations could upset the performance of a remote powering system.

Finally, as a general rule, the 48-volt remote powering standard is recommended. Most capacitor microphones available today call for it, and those that need lower voltages can easily be accommodated through the use of additional dropping resistors, as called out in their published specifications.

CHAPTER V

Microphone Care and Maintenance

INTRODUCTION

It is surprising how many recording and sound engineers of professional standing have little or no knowledge of the basics of microphone care and maintenance. One reason for this is that microphones are relatively cheap and plentiful; people are not inclined to treat them with respect. Too often, the assumption is made that if the microphone works at all, it is working properly.

While modern microphones, especially dynamic types, are quite rugged, they all deserve special care if optimum performance is expected over an extended lifetime.

GENERAL RECOMMENDATIONS

There are a number of rules of good practice which apply to all microphones. We will discuss them below:

1. Protection when not in use. All professional microphones come with some kind of felt or foam-lined box or bag. When the microphone is not in use it should be stored in that container. This may seem a nuisance, but dust—which is everywhere—will slowly find its way into the microphone. In particular, dynamic and ribbon microphones will attract *ferric* dust, minute particles of iron or its oxides which will find their way into the magnetic gaps of those microphones. Just as the serious owner of a camera automatically replaces the lens cover when the camera is not in use, the serious engineer should similarly protect microphones when they are not in use.

2. Cleanliness of working surfaces and areas. Microphones should never be casually laid out on any working surface which has not been carefully cleaned. Carpeted working surfaces may be the worst because the pile tends to trap and hide iron debris which might be subsequently attracted to dynamic or ribbon microphones.

New microphones are often unpacked and checked out in the maintenance department—and this is likely to be the worst possible environment, due to the likelihood of metallic debris. It is best to set aside a clean area where all microphone check-out and maintenance is to be done.

3. Control of Moisture. Today's studios are virtually guaranteed of controlled humidity as a result of air conditioning. And, modern microphones are much less susceptible to moisture than were their predecessors. Older capacitor microphones, with their higher polarizing voltages, were far more susceptible to leakage of the polarizing voltage under conditions of excess humidity. For this reason, many studios in past years carefully monitored humidity conditions and kept their capacitor microphones in a *desiccator* when not in use. (A desiccator is an air-tight enclosure containing a material which attracts moisture.)

By far, the greatest potential for damage due to moisture results from close-up usage by vocalists. The culprits are both condensation of the breath as well as saliva on the mesh and screen which protect the microphone. Over periods of time, the moisture-laden parts of the microphone collect dust particles which become affixed and clog the openings in the mesh. The performance of any microphone depends upon the precise acoustical characteristics of all the openings and passageways to the diaphragm. If these become clogged, the acoustical impedance will vary from that of the original design, and the frequency response of the microphone will deteriorate.

The most difficult task faced by an engineer is that of teaching a performer how to use a microphone. The habits of most performers are thoroughly ingrained, and many of them will resent being taught how to use a microphone. Caution may be the best policy, and every attempt should be made to teach by example. Of course, close positioning of the microphone to the mouth is essential in many applications, and the best that can be done is to make sure that the microphone is examined and wiped with a damp, lint-free cloth after any rigorous usage.

4. Protection from Mechanical Shock. It goes without saying that all microphones should be handled gently and never dropped. Smaller models are remarkably robust, but the larger models require more careful handling. The most fragile of all are older ribbon microphones. The ribbon element itself is delicate, and its natural sub-sonic resonance frequency renders it especially susceptible to shock and low-frequency wind blasts. Ribbons have been known to be deformed by large blasts of breath; even moving a ribbon microphone swiftly on a boom can result in some deformation of the ribbon!

Because it must be fairly open to the air in order to exhibit flat frequency response, the ribbon is also the most susceptible to ferric debris build-up. For these reasons, ribbon microphones have fallen into general disfavor; they are rarely encountered outside of controlled studio environments.

Shock mounts are available for many microphone models. Their primary use is for isolation from floor-borne noise, but additionally they serve to protect the microphone from inadvertent mechanical shock. They should be used whenever possible.

PERIODIC CHECKING OF MICROPHONE RESPONSE

If microphones have been in use for several years, it is wise to check out their frequency response. Short of sending them back to the manufacturer, there are two field tests which will give a good indication of performance. Figure 5-1 shows a method in which a microphone is compared with another of the same model which is known to be in proper working order. Usually, a new microphone, perhaps one borrowed from a dealer, will suffice for this comparison.

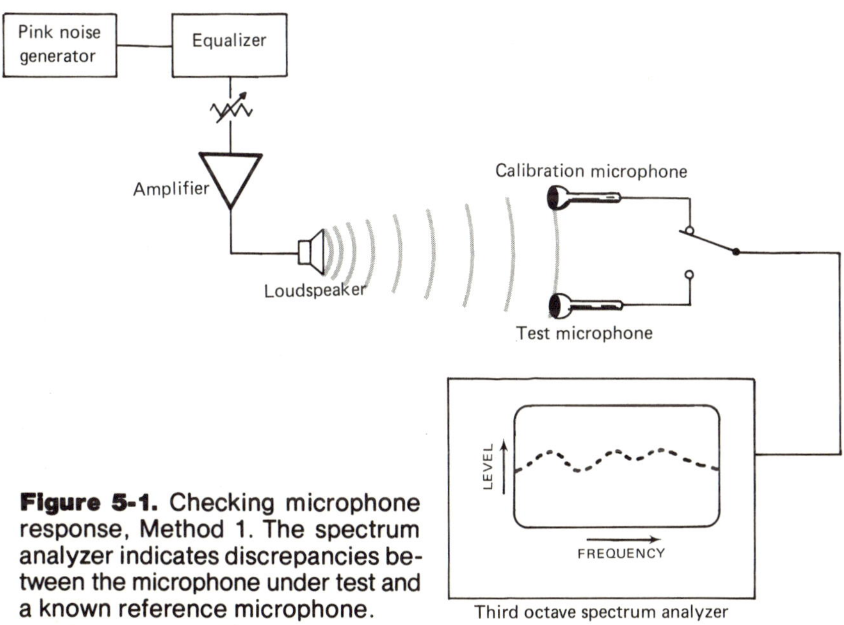

Figure 5-1. Checking microphone response, Method 1. The spectrum analyzer indicates discrepancies between the microphone under test and a known reference microphone.

The test should be made in a fairly dry acoustical environment, and the major axes of both microphones should be precisely aimed toward the loudspeaker. The set-up should be as symmetrical as possible. Use as small a loudspeaker as possible consistent with wide-range response. (If one-third octave equalization is handy, a small five-inch loudspeaker will work well if its response—as measured by the reference microphone—can be made smooth and extended as observed on a real-time spectrum analyzer.)

In making the comparison between the reference and test microphones, look for substantial variations between them, especially at the frequency extremes. Small variations of 1-1½ dB may be due to a slight difference between the positions of the microphones, and of course all microphones have frequency response tolerances which must be taken into account in interpreting the data. This information is normally part of the microphone's data sheet.

Another method of checking the response is shown in Figure 5-2. It is easier to implement than the previous method but is not as precise. It will, however, serve to isolate microphones with serious frequency response problems. In this test, the reference and test microphones are placed together as closely as possible, and a speaker, at a distance of about one meter (three feet) and equidistant from both microphones, talks normally. With one microphone set for normal peaks on a VU meter, raise the level of the other microphone. When the two levels are equal, there should be a drastic reduction in output due to cancellation as provided by the reverse polarity con-

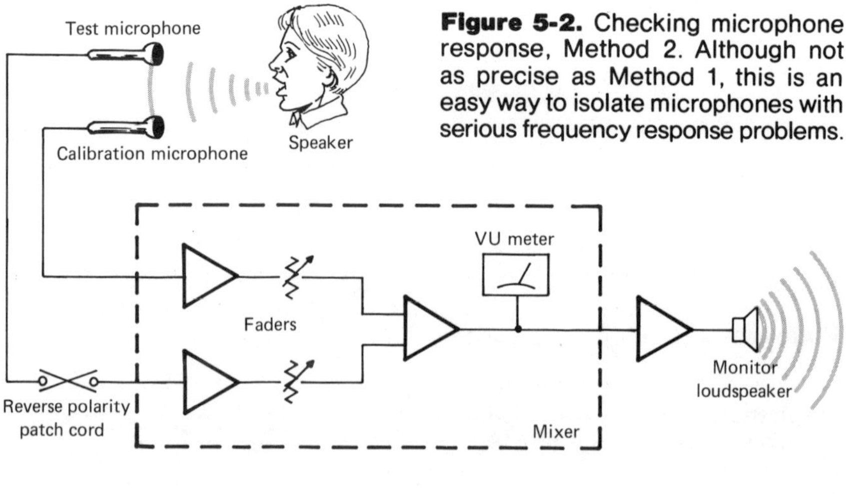

Figure 5-2. Checking microphone response, Method 2. Although not as precise as Method 1, this is an easy way to isolate microphones with serious frequency response problems.

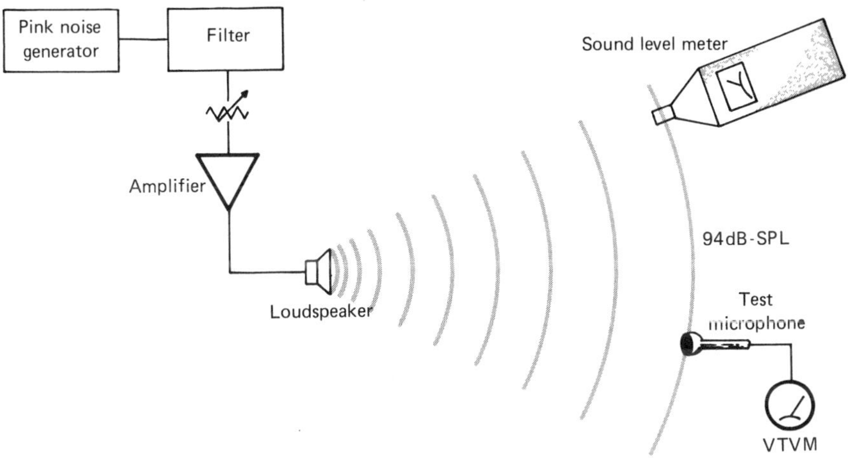

Figure 5-3. Checking microphone sensitivity. With a sound pressure level of 94 dB established, the output voltage of the microphone under test is measured.

nection. The peak level as read on the VU meter should be at least 12-15 dB lower, and the sound, as heard over the monitor loudspeaker, should be reduced in level—but of the same basic sound quality. If it is boomy or excessively sibillant, there is a good chance that the two microphones are not cancelling due to substantial response differences between them. If this test indicates a problem microphone, it should be subjected to a more rigorous test.

Finally, where there is a desire or a need for optimum performance, microphones should be sent back to their manufacturer on a rotating basis, for precise check-out.

CHECKING MICROPHONE SENSITIVITY

Variations in sensitivity between samples of the same microphone model are rarely encountered. Substantial differences in output, when they do occur, are usually in conjunction with some other anomalous behavior, such as gross frequency response difference in dynamic and ribbon types or electrical malfunction in capacitor types. (A trivial case is the 6-dB difference in level due to different strapping of the output transformer, if one is present.)

Figure 5-3 shows a relatively simple set-up for checking microphone sensitivity. Using a sound level meter known to be in proper calibration, establish a sound field of 94 dB-SPL (one pascal) at some fixed distance from a loudspeaker. The loudspeaker's response should be shaped by a one-octave bandpass filter centered at 1 kHz. The test microphone is then placed at exactly the same position occupied by the meter's microphone, and its open-circuit output voltage measured with a VTVM set to the proper scale. By the methods explained in Chapter III, the open-circuit voltage reading may be converted to the format presented in the microphone's published specifications.

REPAIRING MICROPHONES

Outside of minor repair of connectors and fasteners, no attempt should be made to do any repair work on the inner parts of a microphone. With some skill, ferric debris which has been attracted to the outer surface of a dynamic microphone's diaphragm can be removed. A similar pile-up of debris in the gap between the ribbon and pole pieces of a ribbon microphone is virtually impossible to clean without damage to the ribbon; it is best to replace the ribbon, cleaning the pole pieces in the course of doing this.

The effect of ferric debris on the performance of a dynamic microphone will be deterioration of frequency response, with some tendency for buzzing or rattling sounds if the pile-up of debris is especially bad. In a ribbon microphone, the effect of iron filings bridging the gap between pole piece and ribbon will invariably result in buzzing effects.

If the mesh screen surrounding a microphone is loose, it may rattle when subjected to high-level sound fields, especially at low frequencies. Gentle tapping with the finger will usually identify this, and the repair may be well within the scope of a careful maintenance technician.

CABLE INTEGRITY

Microphone cables are subject to continuing abuse in most applications, and a cable testing assembly is essential in any well-run studio organization. Details of a cable testing assembly are shown in Figure 5-4.

This test fixture detects continuity between leads and determines the degree of leakage paths between adjacent connectors.

One of the best ways to isolate a troublesome cable is to connect it between a capacitor microphone and a console equipped for remote powering. The gain over the monitor loudspeakers is then turned up fairly high, and the sound carefully monitored, while a technician

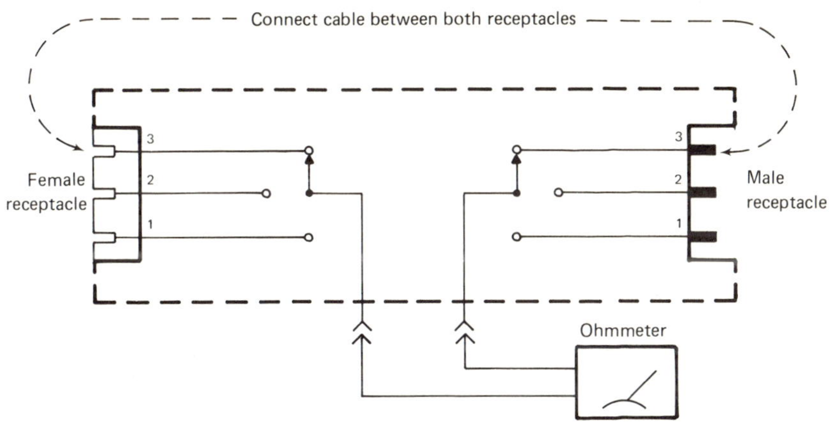

Figure 5-4. Checking microphone cables. For continuity check, set ohmmeter to low scale (R × 1) and read between 1-1, 2-2 and 3-3. Readings should be less than 1 ohm. For leakage check, set ohmmeter to high scale (R × 100k) and read between 1-2, 1-3 and 2-3. If cable is in good repair, no deflection should be observed.

exercises the cable. Any disruptions of the remote powering path will be immediately spotted. In particular, bruised or worn parts of the cable should be flexed, and the portion nearest the connectors should be flexed and tugged on.

ELECTRICAL NOISES

Outside of routine battery replacement in those capacitor microphones which have them, there is little in the way of electronic field maintenance which even a skilled technician has recourse to. Under conditions of temperature or humidity extremes, some capacitor microphones may exhibit deviations from their normal response, but the condition is usually reversible when conditions return to normal. Prolonged exposure to high humidity may result in some increase in noise—typically a sputtering sound—in older capacitor microphones operating with higher polarizing voltages.

If a microphone exhibits intermittent electronic noises, it is a good bet that the problem lies in the preamplifier portion of the capacitor microphone. The best recourse is to send it back to the manufacturer for cleaning and repair.

RF and wireless microphones have their own set of problems, and they are never totally trouble-free. Where wireless operation is essential, the luxury of noise-free response *all of the time* must be foregone.

Most tube-type capacitor microphones encountered today are relics of the past and are prone to all kinds of problems out of sheer age. Many recording companies hold on to certain models because of their particular sonic virtues, but there is no denying that current designs exceed them in reliability and in noise performance.

CHAPTER VI

Using Patterns Effectively

INTRODUCTION

One mark of a professional sound engineer is his thorough understanding of microphone patterns from both theoretical and practical points of view. As he learns his art, he comes to the important conclusion that optimization of any reinforcement, recording, or broadcasting situation, usually involves *fewer* rather than *more* microphones. With experience, he learns that most microphone pickup patterns are only approximations, to a greater or lesser degree, of their nominal patterns, and he must understand that the optimization of his microphone array depends upon a knowledge and appreciation of these departures from the theoretical.

THE CARDIOID FAMILY

Normal cardioid pattern usage is shown in Figure 6-1. The best use of its pattern can be made in fairly dry surroundings where it can be precisely placed with respect to the primary sound (1), and undesired sounds (2), located at 180 degrees.

In addition to normal polar plots, it is helpful to view the directional response of microphones as shown in Figure 6-2. Not all manufacturers present data in this form, but it is an especially helpful tool in determining how natural the off-axis leakage signals will be. The example shown at Figure 6-2A indicates that off-axis sounds will be picked up smoothly in terms of spectrum—but only down in level, as

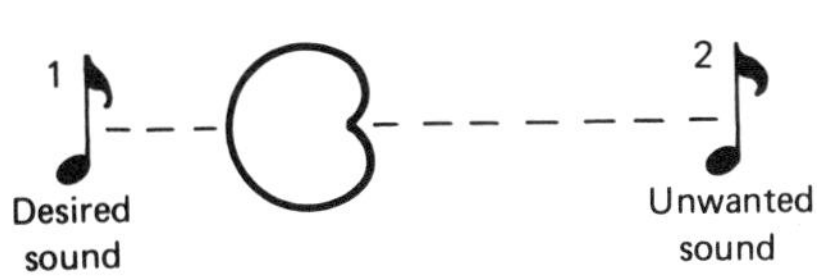

Figure 6-1. Normal use of the cardioid pattern. The desired sound source is located on the microphone's major axis, while unwanted sounds are located 180 degrees off axis.

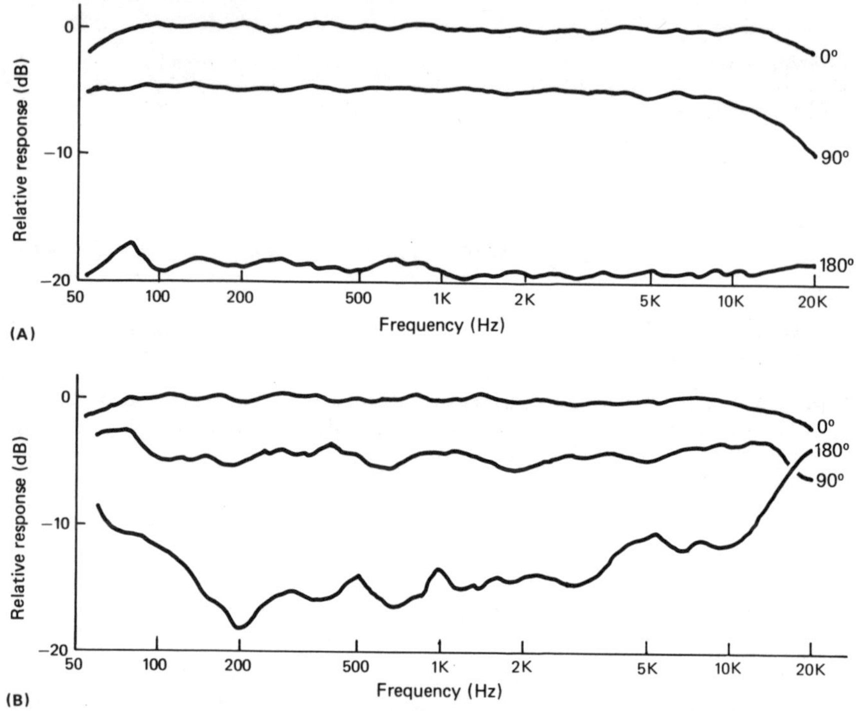

Figure 6-2. Cardioid frequency response versus angle of incidence. The response shown at **(A)** is that of a well-designed microphone, while that shown at **(B)** is typical of many less-expensive models.

expected. At Figure 6-2B, we see an all-too-typical example of a cardioid microphone which, although quite flat on-axis, exhibits peaky response in off-axis directions. With such a microphone, leakage signals (those which are not the primary target of the microphone) will not be natural sounding and may intrude more than they should. When a cardioid microphone is to be used in proximity to a number of instruments, as in a small studio, its off-axis response should be carefully noted.

The hyper-cardioid pattern is normally used in more reverberant environments, because it has more "reach" than the normal cardioid. Referring back to Figure 2-17, we note that the *distance factor* for the hyper-cardioid pattern is twice that of the omni pattern and 1.2 times that of the normal cardioid. The significance of the distance factor is shown in Figure 6-3, where the three patterns (omni, cardioid and hyper-cardioid) are all placed in the same reverberant sound field.

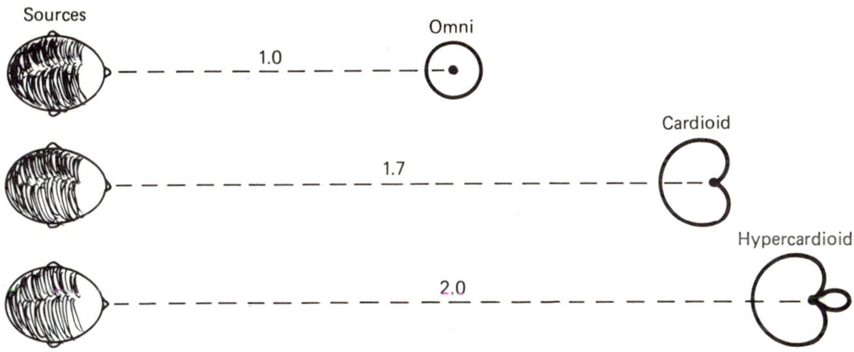

Figure 6-3. Distance factors for omni, cardioid and hyper-cardioid patterns. Directional patterns allow greater working distances than an omni pickup pattern for equivalent immunity to ambient noise.

Under these conditions, sound sources at the distances indicated will be picked up with the same clarity, or direct-to-reverberant ratio.

The hyper-cardioid pattern offers no substantial advantage in a typically dry studio where the reverberant sound field is fairly low. In fact, the presence of the back lobe, which is only 6 dB down from the main lobe, could cause trouble. The greatest utility for the hyper-cardioid will be found in acoustically-live studios or performance environments where the engineer may have only limited flexibility in locating the performers to his best advantage.

Finally, there is the super-cardioid pattern, shown in Figure 6-4. The super-cardioid maximizes the total frontal pick-up (0° ± 90°) as opposed to the total back pickup (180° ± 90°). The overall front-to-back ratio for the super-cardioid is 12 dB. The super-cardioid pattern is therefore very useful when a group of performers is to be picked up by a single microphone. They can be conveniently spread across an arc of, say, 120-to-150 degrees. Under these conditions, the super-cardioid will reject reverberant sound arriving from the back half of the environment by some 12 dB as compared with an omni-directional microphone in the same position. The choice of patterns in this application would depend upon a careful assessment of the various musical factors involved. Specifically, the desired direct-to-reverberant ratio and the required working distance between microphone and ensemble may not be compatible, and the engineer may be forced to work at a greater distance than he would like. Under these conditons, the extra control afforded by the super-cardioid will be appreciated.

Figure 6-4. The super-cardioid pattern provides the maximum frontal pickup relative to total random energy efficiency.

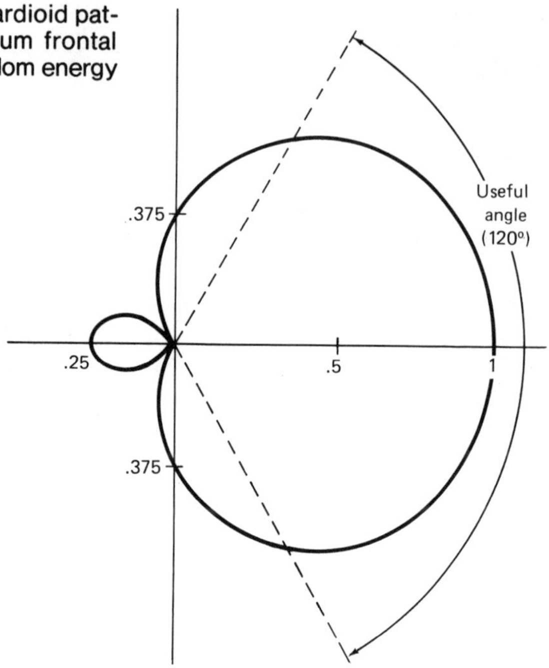

COMPARING CARDIOID MICROPHONES

There are more studio-quality cardioid microphones than any other type, and the inexperienced engineer can easily become confused in his attempts to compare their specifications. In general, more accurate cardioid performance across the frequency band is afforded by those microphones which have only a single pattern. These are invariably of the single-element variety and, as a rule, have their major pickup lobe along the microphone's physical axis, as shown in Figure 6-5A.

Microphones with variable patterns usually have their major lobe at right angles to microphone's physical axis, as shown in Figure 6-5B. The lack of rotational symmetry about the major lobe of many of these models may account for slight irregularities in the patterns at high frequencies due to diffraction effects around the microphone case.

Both polar plots and families of off-axis frequency response curves should be studied and comparisons should be carefully made. The on-axis smoothness as well as extension of high- and low-frequency response should be compared.

Figure 6-5. Axial **(A)** and radial **(B)** directions of maximum sensitivity. *(Gotham Audio Corp. photo.)*

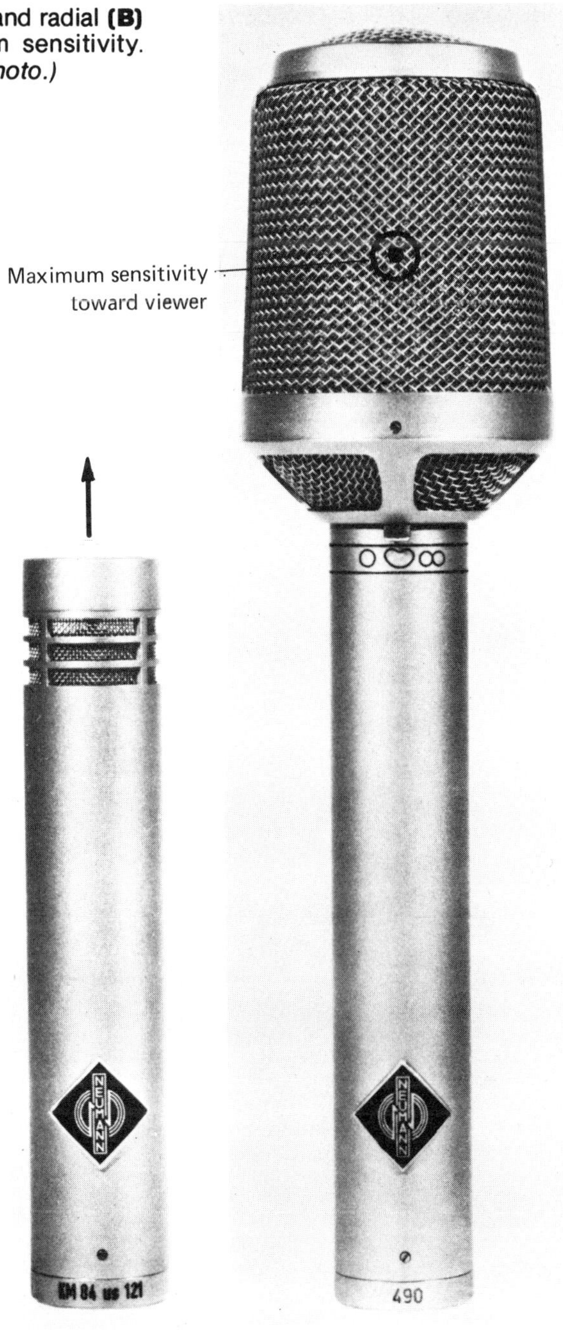

(A) (B)

THE CARDIOID MICROPHONE IN SOUND REINFORCEMENT

The concept of acoustical gain in a reinforcement system will be discussed in detail in a later chapter. At this point, we will only make broad observations on the behavior of reinforcement systems and note those specific conditions under which a cardioid pattern may offer some advantage over an omni-directional microphone. Let us first consider a reinforcement system located outdoors, where the loudspeaker and microphone will be well away from reflective surfaces. The basic elements are shown in Figure 6-6.

Under outdoor conditions, sound pressure falls off approximately 6 dB with each doubling of distance away from the sound source,

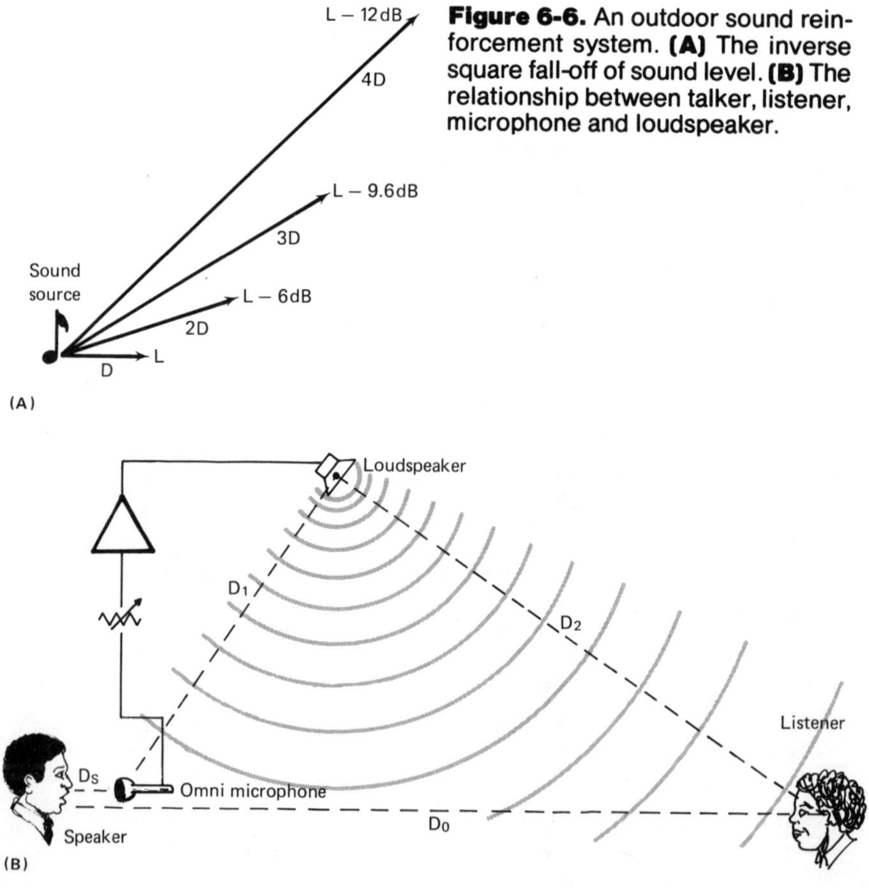

Figure 6-6. An outdoor sound reinforcement system. **(A)** The inverse square fall-off of sound level. **(B)** The relationship between talker, listener, microphone and loudspeaker.

assuming that the effect of reflections can be ignored. This is known as inverse-square loss and is given as

$$\text{Loss (dB)} = 10 \log \left(\frac{r_2}{r_1}\right)^2 = 20 \log \left(\frac{r_2}{r_1}\right)$$

where

r_1 = reference distance
r_2 = measurement distance

In the system shown in Figure 6-6B, both microphone and loudspeaker are assumed to be omni-directional. The gain of the amplifier can, of course, be increased up to some point at which the signal fed back from the loudspeaker through the microphone will cause the system to "howl" or "ring." Under these conditions, the criteria for system stability will no longer exist.

The distances D_S and D_1 can be manipulated to increase the system gain: if we halve D_S or double D_1, we will increase the margin before feedback by 6 dB. We may manipulate D_S and D_1 independently, and the effect on gain will be additive in terms of decibels. The nomograph of Figure 6-7 provides a direct solution to the problem.

As an example, consider D_S to be fixed. At Figure 6-8A, we have doubled D_1 and thus increased the system stability margin by 6 dB. At Figure 6-8B, we have used a cardioid microphone and precisely located it so that the loudspeaker is 90 degrees off its major axis where the microphone's response is 6 dB down. In theory at least, the two situations are the same, and both should accomplish the same 6 dB increase in stability margin.

Again, in theory, the situation shown in Figure 6-8C should increase the stability margin significantly; placing the microphone so that its null angle is pointed toward the loudspeaker would appear to isolate the microphone totally from the influence of the loudspeaker.

Things are not so simple in practice, and it usually turns out that manipulations of D_S and D_1 are more effective than cardioid microphone orientation in increasing the feedback stability margin. As a general rule, no more than 4 or 5 dB improvement in system gain should be expected through the use of directional microphones. There are always reflections present, and, of course, the microphone's polar response will never be ideal.

When we move indoors, the problem is more complex due to reverberation. Reverberation is the result of an ensemble of reflec-

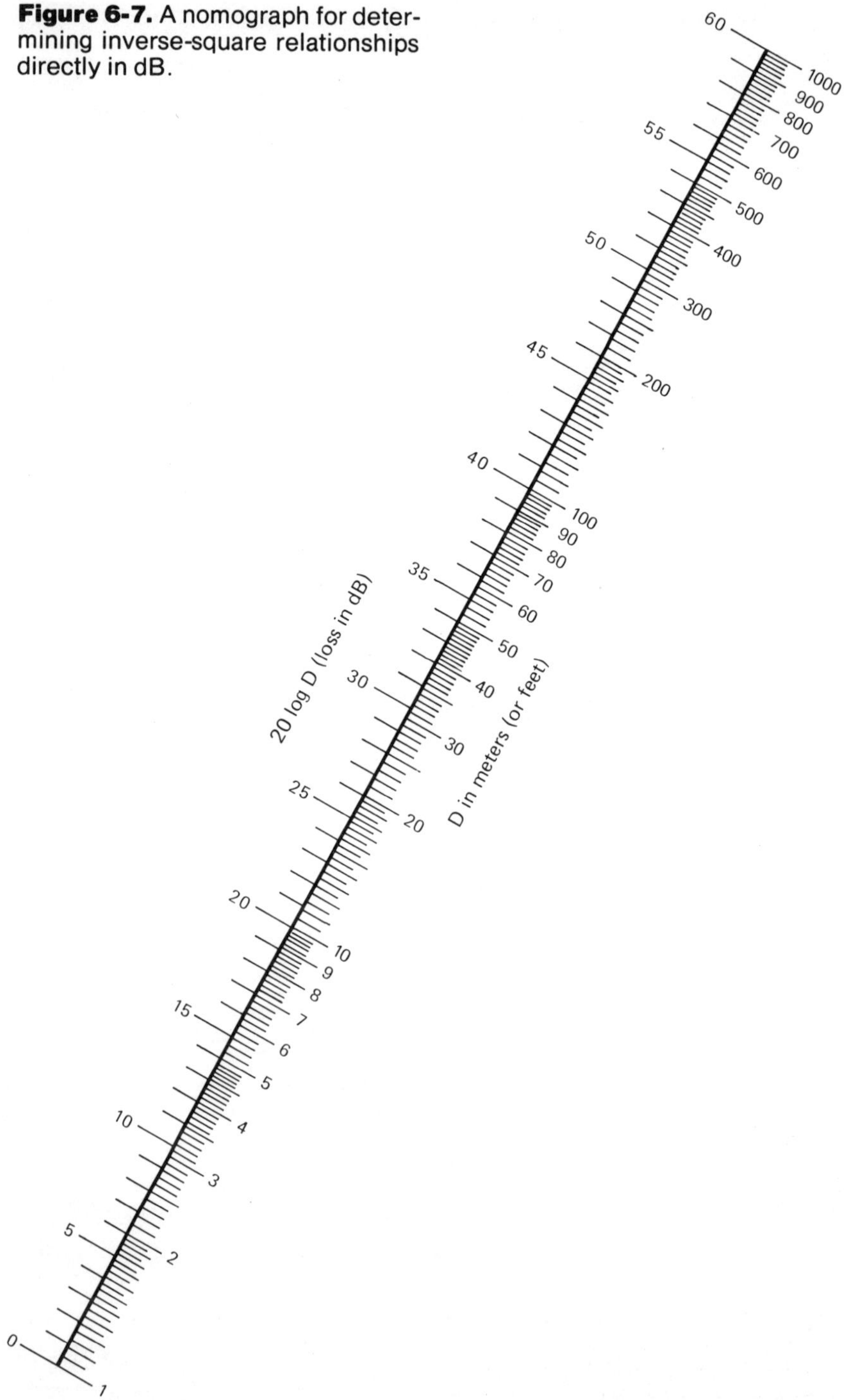

Figure 6-7. A nomograph for determining inverse-square relationships directly in dB.

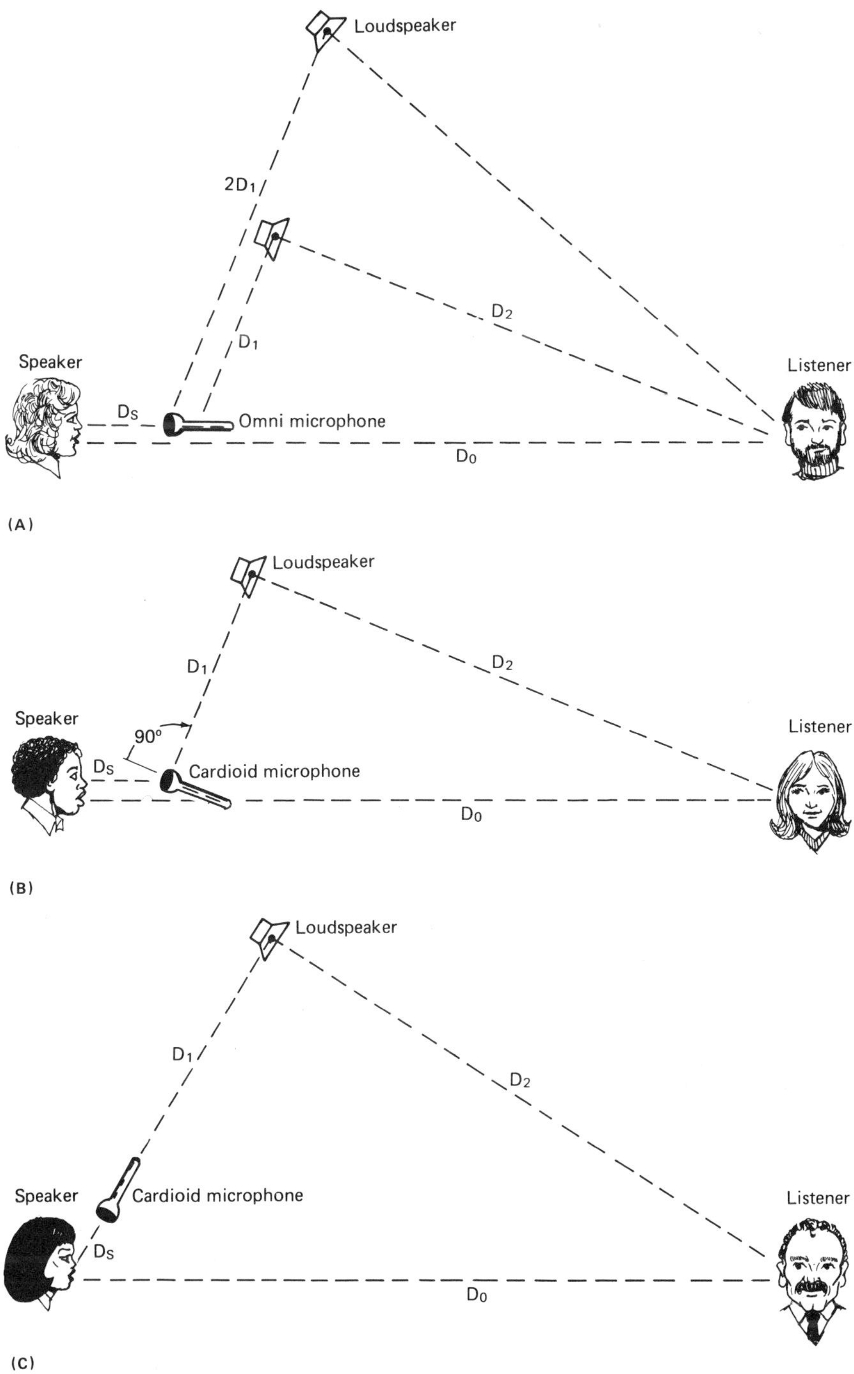

Figure 6-8. Increasing system gain by **(A)** increasing the distance between the loudspeaker and the microphone, **(B)** orienting the microphone's pick-up pattern so that the loudspeaker is 90 degrees off axis, **(C)** optimizing the microphone's orientation so that its null angle faces the loudspeaker.

tions in an enclosed space, and its effect on the attenuation of sound as the observer moves away from a source is as shown in Figure 6-9. Note that for positions fairly close to the sound source, sound is attenuated according to the inverse-square law: 6 dB loss with each doubling of distance. However, at greater distances, the reverberant sound field predominates, and there will be little attenuation beyond a certain distance. This distance is referred to as *critical distance* (D_C) or as the *reverberation radius,* the distance from a sound source where both the direct and reverberant sound fields are equal. In a later chapter we will develop equations for calculating D_C. For the present, we will note in non-analytical terms the effect of the reverberant field on the gain of the reinforcement system.

We noted in the outdoor case that each doubling of D_1 increased the system stability margin by 6 dB. In the indoors case this will be true only up to the point where D_1 is roughly equal to D_C; beyond D_C, any further increase in D_1 will not improve system stability. It seems obvious then, that in an indoor sound reinforcement system, the loudspeaker-to-microphone distance should always be *greater* than D_C. When D_1 exceeds D_C, any sound from the loudspeaker picked up by the microphone will be from the essentially random reverberant sound field.

With indoor systems, a cardioid pattern may offer some advantages over an omni. Referring back to Figure 2-17, we noted that the random energy efficiency (REE) of a cardioid microphone is 0.333;

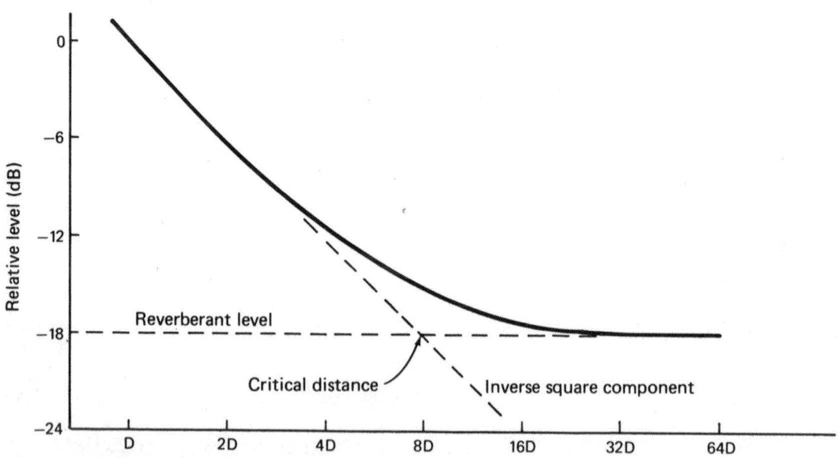

Figure 6-9. Attenuation of sound indoors. The reverberant field becomes prominent at greater distances from the sound source.

this means that it will pick up only one-third as much reverberant energy as an omni when both are used at the same distance for picking up a speaker. The REE of the cardioid translates into an additional 4.8 dB of system gain or feedback stability margin. In the case of the hyper-cardioid pattern, the improvement will be 6 dB.

Of course, for very close microphone placement, there will be more than adequate gain, and a directional microphone will have no advantage over an omni. The omni pattern is, in fact, the preferred one unless there are specific reasons for choosing a directional microphone for sound reinforcement work.

THE BI-DIRECTIONAL (FIGURE-8) MICROPHONE

Because of its relative fragility, the ribbon microphone is rarely encountered outside the studio. Over the years it has had its adherents in broadcast work because of its natural application in dialog pick-up, where the two speakers face each other with the microphone between them. Those bi-directional patterns produced by the Braunmühl-Weber capacitor construction have considerable application in the recording studio and are generally more reliable than the ribbon types.

Some typical uses of the bi-directional pattern are as follows:

1. **Dialog pickup.** As shown in Figure 6-10A, performers can be placed on opposite sides of the pattern, maintaining considerable realism in terms of natural balances. The REE of the pattern is 0.333, and the rejection of reverberant noise is some 4.8 dB over an omnidirectional microphone in the same application.

2. **Speakers with audience.** As shown in Figure 6-10B, the null plane of a bi-directional pattern can be positioned so as to minimize the interference of audience noises with a performer.

3. **Specific instrumental pickup.** Where an instrument has its sound producing portions divided into two sections, a bi-directional pattern can be useful, as shown in Figure 6-10C.

When a bi-directional microphone is not handy, one can be synthesized from a pair of matched cardioids, as shown in Figure 6-11.

When the cardioids are closely placed and connected in opposite polarity, the component patterns (remember the explanation in Chapter II) are combined as shown.

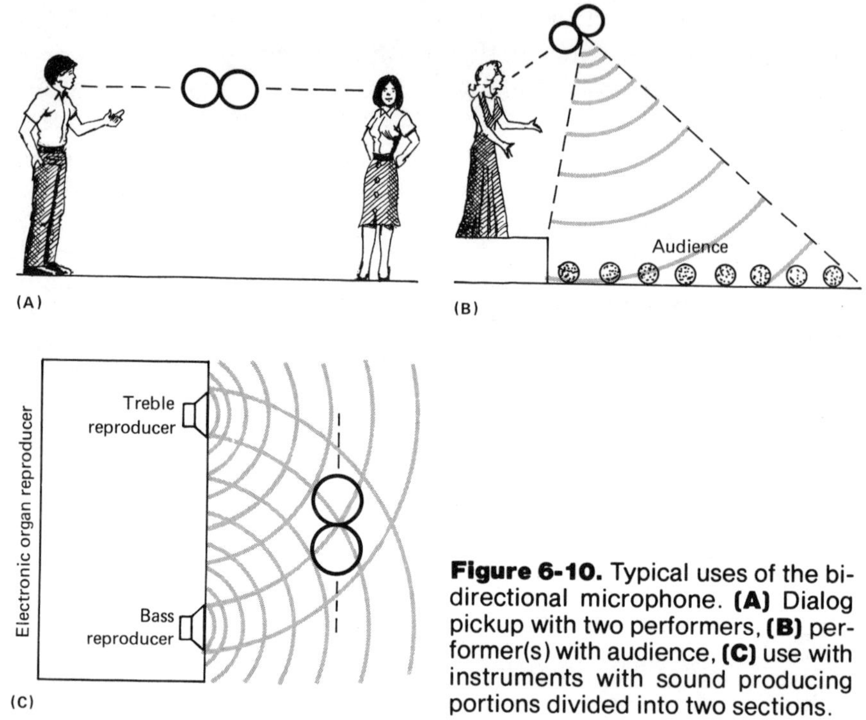

Figure 6-10. Typical uses of the bi-directional microphone. **(A)** Dialog pickup with two performers, **(B)** performer(s) with audience, **(C)** use with instruments with sound producing portions divided into two sections.

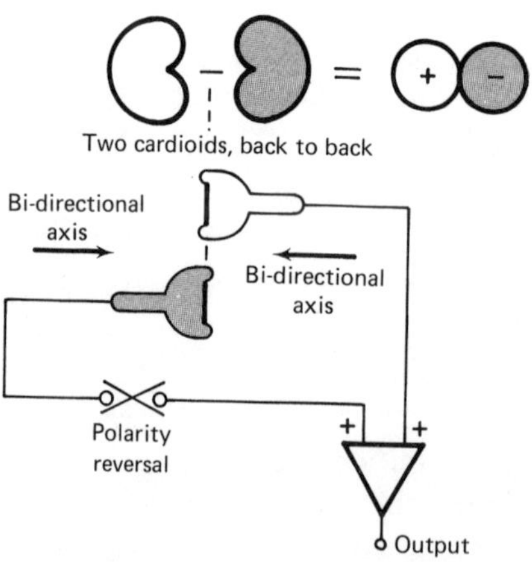

Figure 6-11. A combination of back-to-back cardioid microphones may be used to create a bi-directional response.

The bi-directional pattern is not normally used in sound reinforcement work. The capacitor types are generally too expensive to be subjected to the rigors of reinforcement work, while the ribbon types are too fragile. Also, the proximity effect of a bi-directional microphone is quite pronounced and can be very troublesome in reinforcement work. This topic will be discussed in detail in the following chapter.

The bi-directional pattern has a very special place in the art of stereophonic recording, where a pair of them form the basis of what is called *intensity stereo.* In this application, the precise nature of the microphone's polar pattern results in a startling degree of realism through the creation of accurate phantom images in stereo playback. This topic will be covered in detail in Chapter 10.

THE OMNI-DIRECTIONAL PATTERN

In general, the omni-directional microphone, whether dynamic or capacitor, is a simple and reliable device. However, these microphones depart from omni-directionality at very high frequencies. Because of the finite size of the diaphragm, they will show a preference for zero-degree incidence signals. The larger the diaphragm, the lower in frequency range this preference will be shown. Figure 6-12 shows typical response for signals at 0 degrees, 90 degrees and of random incidence. Care should always be taken that the maximally sensitive high-frequency lobe of an omni-directional microphone is pointed at the sound source.

For those omnis which have their maximum sensitivity along the physical axis of the microphone, the high-frequency response will be modified as shown in Figure 6-13A. For the Braunmühl-Weber type of capacitor microphone, the high-frequency pattern will be modified

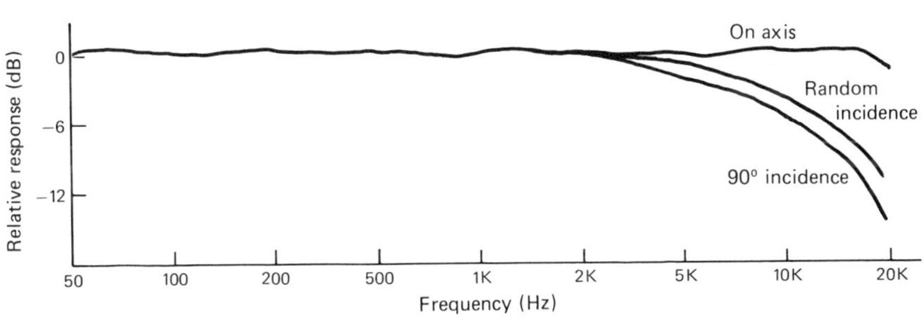

Figure 6-12. Omni-directional microphone response.

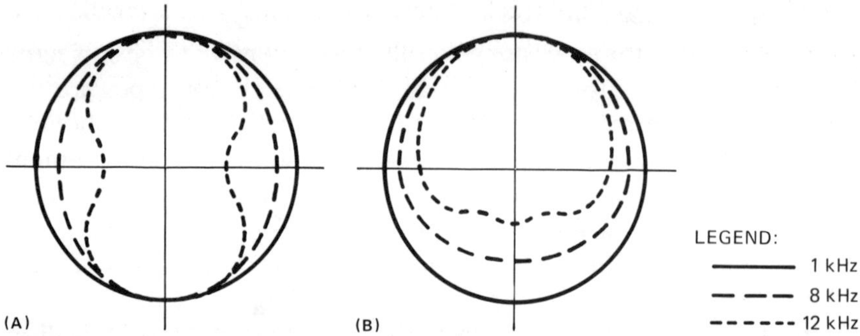

Figure 6-13. The high-frequency polar response of **(A)** radially sensitive and **(B),** axially sensitive microphones.

as shown at Figure 6-13B. These microphones have their maximally sensitive directions at right angles to the microphones axis, and the fore-aft symmetry of the design results in a bilaterally symmetrical pattern.

In this regard, single-element omnis tend to exhibit cardioid-like behavior at high frequencies, while the Braunmühl-Weber design shows a tendency toward bi-directional response at high frequencies.

It is easier to make a good, extended-range omni microphone than either a cardioid or bi-directional. As a rule, the omni will exhibit a minimum of coloration and should normally be first choice in all situations where a specific directional pattern is not called for. Sound reinforcement problems are almost always minimized in situations of adequate potential gain when an omni is chosen. Their freedom from proximity effect makes them especially useful for close-in placement to instruments in the studio.

INVERSE-SQUARE LAW IN THE STUDIO

With omni-directional microphones, distances rather than orientations are paramount. The goal in so much of modern pop/rock recording is to isolate the various musical ingredients so that they may be stored on individual tracks of a multi-channel recorder for subsequent mixing down to a final format.

It is important that the studio engineer understand the effects of those distance variables which contribute to separation. Let us examine a hypothetical studio situation, as shown in Figure 6-14A. Here, we have a brass instrument, 1, and a woodwind instrument, 2, each to be recorded on its individual track. There will generally be a problem of the louder brass instrument leaking into the woodwind microphone. By inverse-square law, the level of source 1 at the microphone intended for source 2 will be, relative to the level of source 2 at the same microphone,

$$\text{Level} = 20 \log (D_2/D_1)$$

For typical distances, say, $D_2 = 1\text{m}$ (3 ft.) and $D_1 = 3$ m (9.5 ft.), the level of source 1 will be down some 9.6 dB at the woodwind microphone relative to source 2. All would be well if both instruments were playing at the same level. This is not likely, and we can normally expect a brass instrument to play, at the least, a good 10 dB louder than a typical woodwind. If this is the case, then the 9.6 dB advantage that source 2 has at its own microphone will be swamped out, and both sources *will be picked up about equally!*

Let us consider some solutions to this problem. At 6-14B, we have *halved* the distance D_2, picking up another 6 dB in favor of source 2.

$$\text{Level} = 20 \log (D_2/2D_1) = 15.6 \text{ dB.}$$

Assuming that the studio is sufficiently inert acoustically, we can double distance D_1, achieving essentially the same result as above (shown at 6-14C).

Of these two solutions, halving D_2 is easier, and more quickly accomplished. It may or may not be sufficient, depending on many other factors. Any further shortening of D_2 may cause problems; if

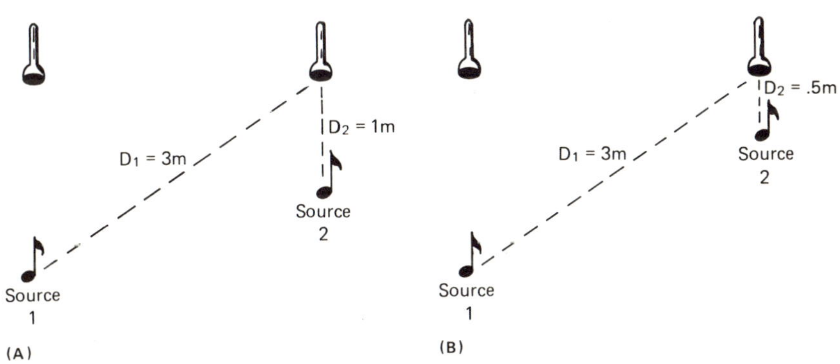

Figure 6-14. The inverse square law in the studio. Distance ratios of **(A)** 3-to-1 and **(B)** 6-to-1 between two sources and one microphone.

the microphone is too close to the instrument, then even minor changes in distance between microphone and instrument caused by natural movements of the player will result in noticeable shifts in level.

A further remedy would be to use a cardioid microphone on source 2, with its null direction aimed at source 1. Additional remedies call for the use of acoustical isolators around source 1 to contain it to a higher degree, a technique to be discussed in a later chapter. All three remedies would probably yield satisfactory results.

INSTRUMENTATION MICROPHONES

Instrumentation microphones are invariably of the omni-directional capacitor type. They are the class of microphones used in all serious acoustical measurement—and of course are the one by which all commercial microphones are measured. Figure 6-15 shows the response of Brüel & Kjaer instrumentation microphones of nominal quarter-, half-, and one-inch diameters. With each succeeding halving of diaphragm diameter, note that the frequency region where response anomalies accour is shifted up one octave.

The large diameter designs are used for low-noise measurements, while the smaller diaphragms are used for relatively high-level, high-frequency measurements.

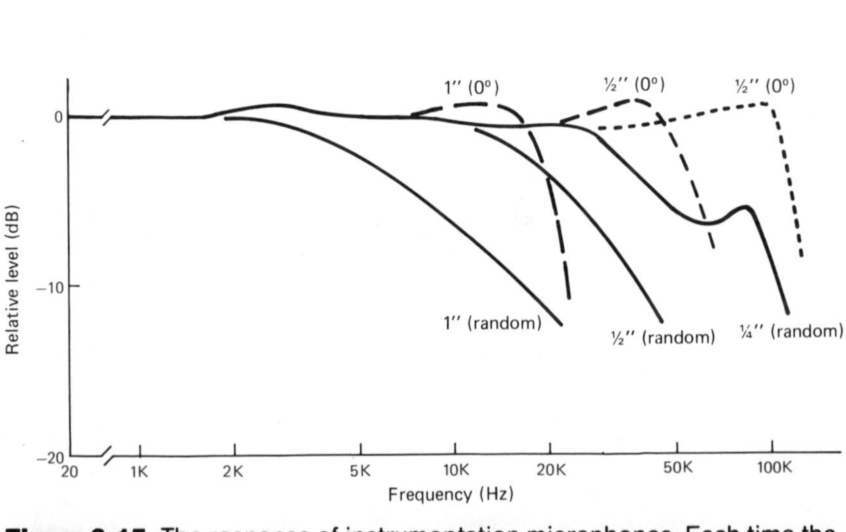

Figure 6-15. The response of instrumentation microphones. Each time the size of the diaphragm is halved, the frequency response limits are doubled.

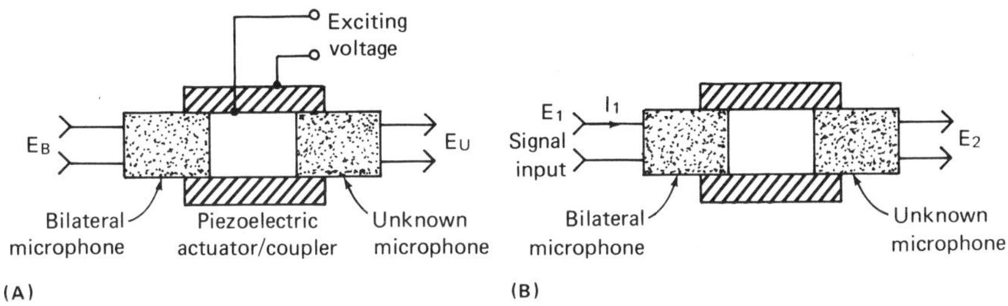

Figure 6-16. Reciprocity calibration of microphones. Determining the ratio **(A)** and product **(B)** of microphone sensitivities.

CALIBRATION OF MICROPHONES

Reciprocity calibration is normally used to determine the frequency response and sensitivity of test microphones. Reciprocity calibration is based on the property of certain transducers to act both as microphones and as miniature loudspeakers. Such transducers are referred to as *bi-lateral,* and their response—when analyzed as linear circuit elements—obeys the law of reciprocity and enables these microphones, in a sense, to become self-calibrating. Both capacitor microphones and those operating on the principle of magnetic induction can be used for calibration purposes.

In the most general form, reciprocity calibration is performed in two steps, as shown in Figure 6-16. First, both the unknown microphone and the bi-lateral test microphone are excited by a third transducer of unknown characteristics. The purpose of this procedure is to establish the *ratio* of the sensitivities of the bi-lateral and unknown microphones. Only the ratio, and not absolute values are sought in this step.

The next step in the procedure is to plot the response of the unknown microphone by driving the bi-lateral microphone as a miniature loudspeaker. In this step, we can determine the product of the two sensitivities, taking into account both electrical and acoustical equivalent circuit elements. The sensitivity of either microphone can then be determined through simple algebraic manipulation of the ratio and product of sensitivities.

Reciprocity calibration is a rather critcal test procedure involving precise electrical measurements and calculation of acoustical impedance. As a more routine check of microphone sensitivity, a *piston-*

phone can be used. A pistonphone is a mechanically-actuated driver producing a known sound pressure level in the frequency range of 350-400 Hz. Its accuracy is usually of the order of ± 0.1 dB. If the assumption can be made that the frequency response of a test microphone is within normal tolerance, then the pistonphone will be sufficient for overall acoustical calibration of a measuring system.

SUMMARY

In this chapter, we have covered the properties of the basic patterns and some of the rules governing their choice in a variety of applications. Other important aspects governing the choice of microphones deal with proximity effects and the effects of nearby reflecting surfaces. As the student progresses on through the book, he will appreciate the degree to which all these factors are interrelated.

CHAPTER VII

Proximity and Distance Effects

CAUSE OF PROXIMITY EFFECT

When a cardioid or bi-directional microphone is placed close to a sound source, its bass response rises relative to its mid- and high-frequency response. The effect has been used for years to good advantage by vocal soloists who appreciate the extra fullness and richness in vocal timbre which it provides. Pressure (omni-directional) microphones do not exhibit this effect, for reasons which we will soon appreciate.

In Chapter 1 we discussed the operation of the ribbon microphone, and at this point we ask the reader to review that discussion. In particular, note in Figure 1-4B that the force acting on the ribbon is due to a pressure gradient which depends on the frequency of the signal. That force rises with frequency at the rate of 6 dB-per-octave. When the microphone is at some distance from the source, then that force is the only one operant. However, at close working distances, there will be an additional force, and its derivation is shown in Figure 7-1.

Since the ribbon, or any other pressure gradient microphone senses pressure at two points, we must take into account the difference between these points due to inverse square law, as shown at Figure 7-1A. The pressure gradient due to inverse-square law is constant with respect to frequency and is equal to

$$\text{Difference} = (D_1/D_2)^2.$$

The omni microphone senses pressure at only one point; therefore this effect does not exist with that type.

As shown in Figure 7-1B, the inverse-square force is dominant at low frequencies, while the force due to the phase relationship between the two sensing points is dominant at high frequencies. When the forces act on the mass-controlled ribbon, the net output of the microphone is as shown at Figure 7-1C.

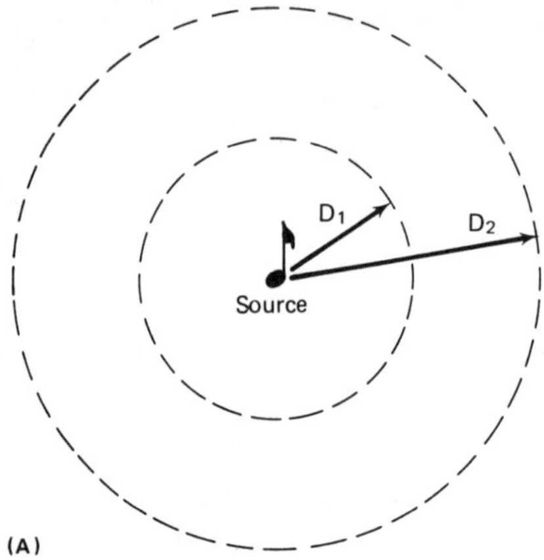

Figure 7-1. The proximity effect in pressure-gradient microphones. **(A)** Sound pressure levels are proportional to the square of the distance from the source. **(B)** The combination of inverse-square and frequency-dependent forces on a ribbon. **(C)** The electrical output of a ribbon microphone, due to the forces shown at **(B)**.

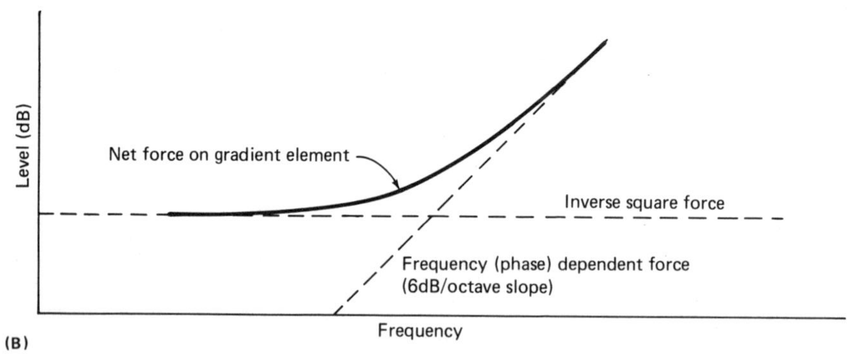

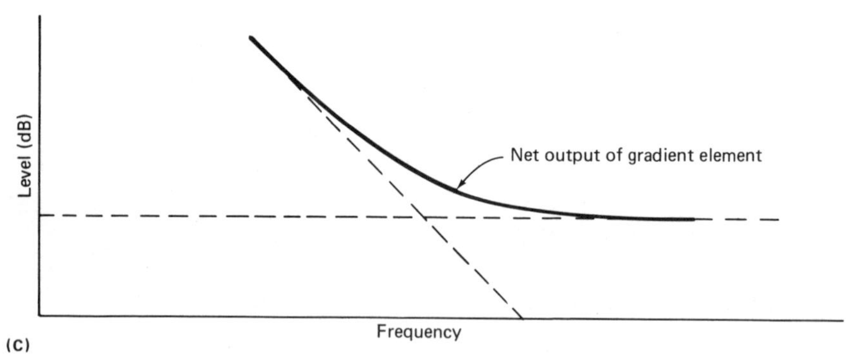

In mathematical terms, the magnitude of the proximity effect is expressed as

$$\frac{\sqrt{1 + (kr)^2}}{kr}$$

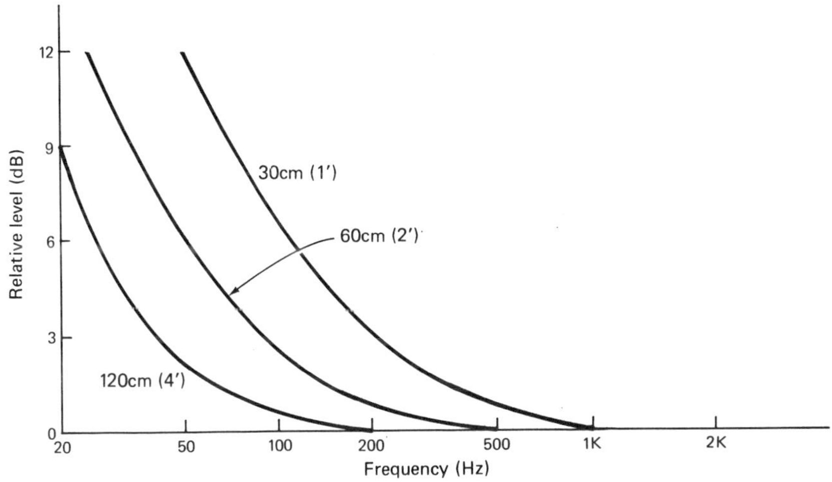

Figure 7-2. Calculated plots of proximity effect for a bi-directional microphone.

where:

r = distance from source,

$k = \frac{2\pi f}{c}$,

c = velocity of sound, and

f = frequency.

Plots of this expression are given in Figure 7-2 for several values of r.

For a cardioid microphone, the proximity effect will be less than that for the bi-directional gradient microphone. Since the cardioid consists of omni- and bi-directional components in equal amounts, we can expect the proximity effect to be roughly 6 dB less than for the bi-directional at the same working distance.

In addition, the degree of proximity effect exhibited by the uni-directional microphone will depend upon the bearing angle, as shown in Figure 7-3. Note that for an angle of 90 degrees there is no proximity effect, since at that angle only the omni component is operating. Another consequence of the proximity effect in a cardioid microphone is that at low frequencies and at short working distances it exhibits a pronounced bi-directional response.

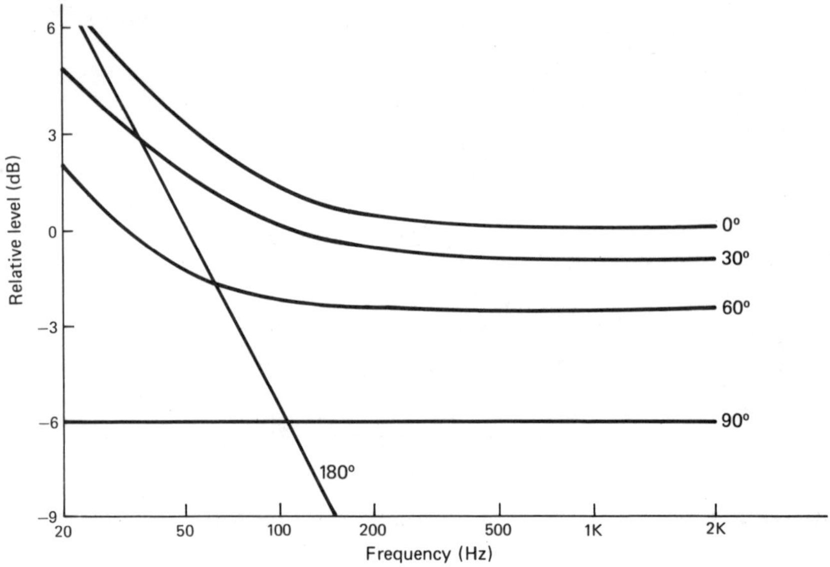

Figure 7-3. Calculated plots of proximity effect for a uni-directional microphone at 60 cm (2 ft.).

PROXIMITY EFFECT IN TYPICAL MICROPHONES

A number of manufacturers publish curves showing proximity effect for several working distances. In the three examples to follow, we will examine microphones intended for close usage. Their low frequency response at large distances has been rolled off purposely so that their net proximity effects will be somewhat less than that shown in the calculated data of Figure 7-3. For a dynamic cardioid microphone of conventional construction, the curves shown in Figure 7-4 are typical.

Capacitor cardioid microphones exhibit similar curves, as shown in Figure 7-5.

The Variable-D dynamic cardioid design discussed in Chapter 2 exhibits significantly less proximity effect, as shown in Figure 7-6. This design has a number of frequency-selective paths to the rear of the diaphragm, and the effect of these is to alter in a significant way the balance between the inverse-square and phase-induced gradients acting on the diaphragm at low frequencies. As a result of this, the net proximity effect is relatively small as compared with standard cardioid dynamic microphones.

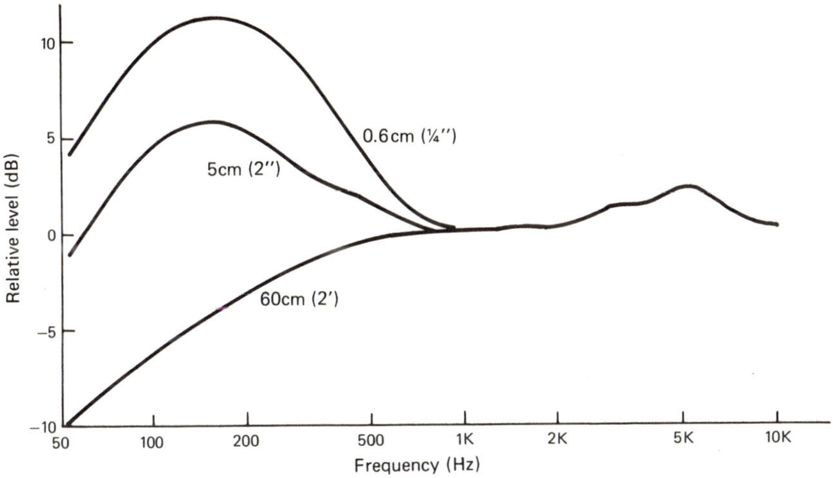

Figure 7-4. Proximity effects in a typical dynamic cardioid microphone.

USES OF THE PROXIMITY EFFECT

Singers have long appreciated the increase in fullness which accompanies close usage of cardioid and bi-directional microphones. The "bigger-than-life" effect has actually become a part of the sound recording and reinforcement arts, and directional microphones are

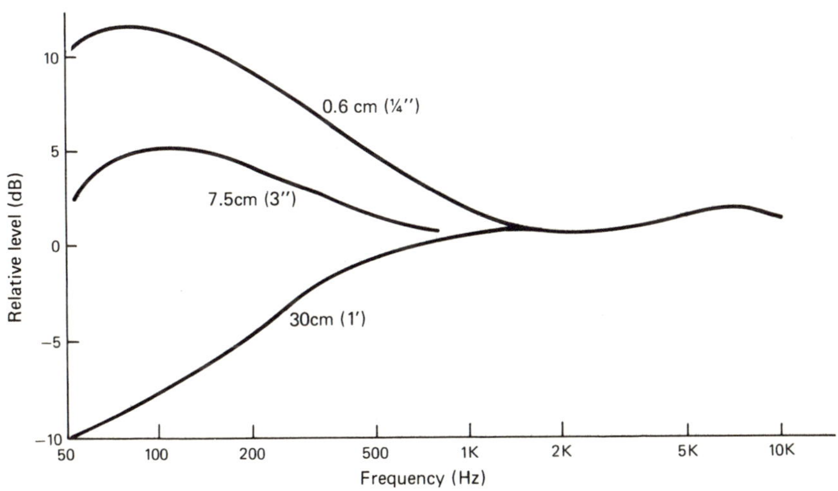

Figure 7-5. Proximity effects in a typical capacitor cardioid microphone.

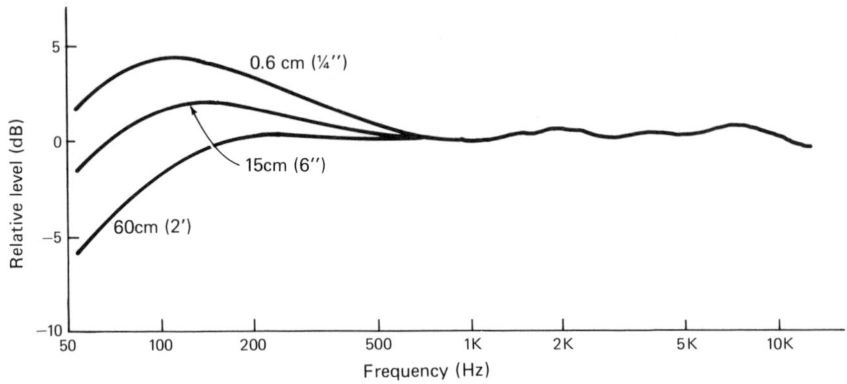

Figure 7-6. Proximity effects in a typical Variable-D microphone.

often chosen for hand-held use by singers. It is very important in such hand-held usage of directional microphones for vocal pickup that the microphone-to-mouth distance be maintained as fixed as possible. If this distance is allowed to vary, then not only will the level rise and fall, but the degree of proximity effect will vary as well. Good engineering practice calls for the use of a compressor or limiter at some point in the vocal channel, but this may not be enough. The performer needs to be taught the rudiments of proper microphone handling.

The proximity effect is not generally desired for speech because of its distinctly unnatural sound. Many directional microphones incorporate a speech/music switch, which inserts a low-frequency roll-off in the speech position, partially compensating for the proximity effect. Details of this are shown in Figure 7-7.

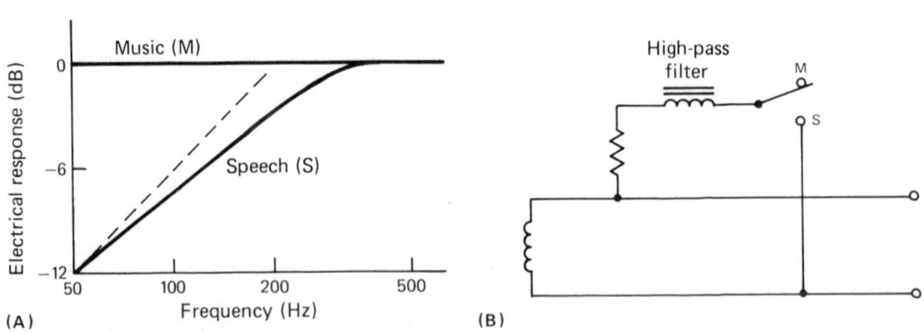

Figure 7-7. The effect of a speech/music switch **(A),** and a typical schematic for the speech/music filter **(B).**

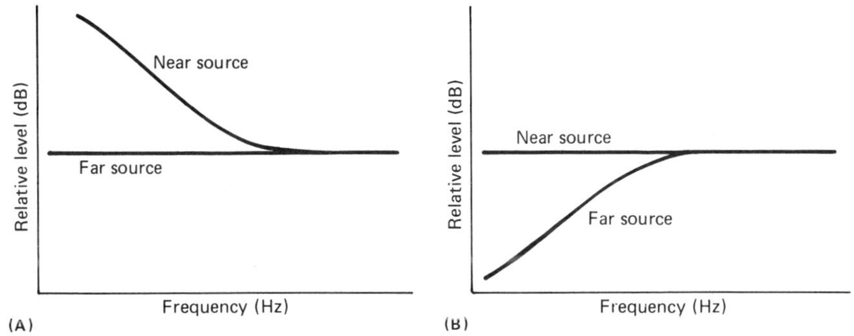

Figure 7-8. The noise cancelling aspect of the proximity effect. **(A)** normal proximity effect; **(B)** microphone response with low frequencies rolled off.

In sound reinforcement work, the proximity effect can work to some degree as a noise-cancelling microphone, as shown in Figure 7-8. If the response is equalized for flat response of the near source, then distant sources will be rolled off, significantly reducing their annoyance level.

THE NOISE-CANCELLING MICROPHONE

A noise-cancelling microphone is one optimized for close placement to a speaker in noisy environments. Ideally, its response to far and close sources would be as shown in Figure 7-9A. Since the normal spectrum of noise is more-or-less limited to mid and low frequencies, the noise-cancelling microphone will reject those components quite effectively. Actually, the noise-cancelling microphone is a pressure-gradient device which has been equalized so that near sounds, boosted by the proximity effect, will be flat, while far sounds will be rolled off at low frequencies.

A reasonably good noise-cancelling microphone may be constructed as shown in Figure 7-9C. Here, a pair of dynamic omni microphones are taped together and connected out of phase. The speaker talks directly into one of the microphones as shown. The voice level at the near microphone is of course much greater than at the far microphone, and the output will be predominantly due to the near microphone. However, distant interfering sources will be received by each microphone about equally, and they will thus be effectively cancelled.

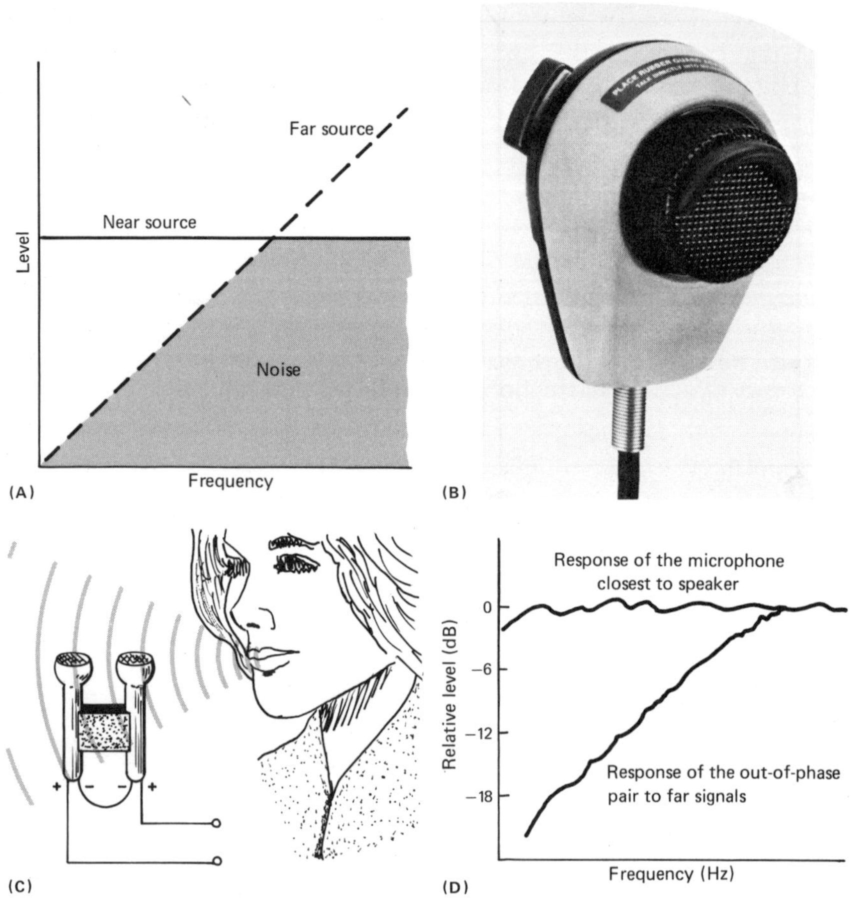

Figure 7-9. The principle of the noise cancelling microphone. **(A)** The response to near and far sound sources; **(B)** a typical noise cancelling microphone *(Shure Bros. photo)*; **(C)** out-of-phase wiring of two omni-directional microphones to produce a noise-cancelling effect; and **(D)** the effective response of the pair.

DIFFRACTION EFFECTS

At long wavelengths (low frequencies), sound tends to bend or *diffract* around obstacles, while at short wavelengths (high frequencies), the obstacle tends to produce a shadow. As shown in Figure 7-10, a low-frequency sound arriving on-axis or off-axis at an omni microphone will be the same. However, a high-frequency sound will be stronger on-axis than it will be off-axis. The reasons for this have to do with sound reflections which take place at the edges of the microphone housing as well as the fact that the microphone may intercept a sizeable portion of a wavelength for an off-axis signal. The variations in high-frequency polar response discussed in Chapter 6 are due to such effects as these.

If a pressure element of the order of 2-3 cm (1 inch) in diameter is placed in a free field at zero-degrees *incidence*, there will be a broad rise of about 6 dB in the 10 kHz range. The microphone may be designed so that its zero-degree incidence response is flat; in that case the random incidence response would roll off as shown in Figure 7-11A. Alternatively, if the random incidence response is maintained flat, then the zero degree incidence response will be peaked, as shown in Figure 7-11B.

Like the proximity effect, high-frequency peaks due to diffraction

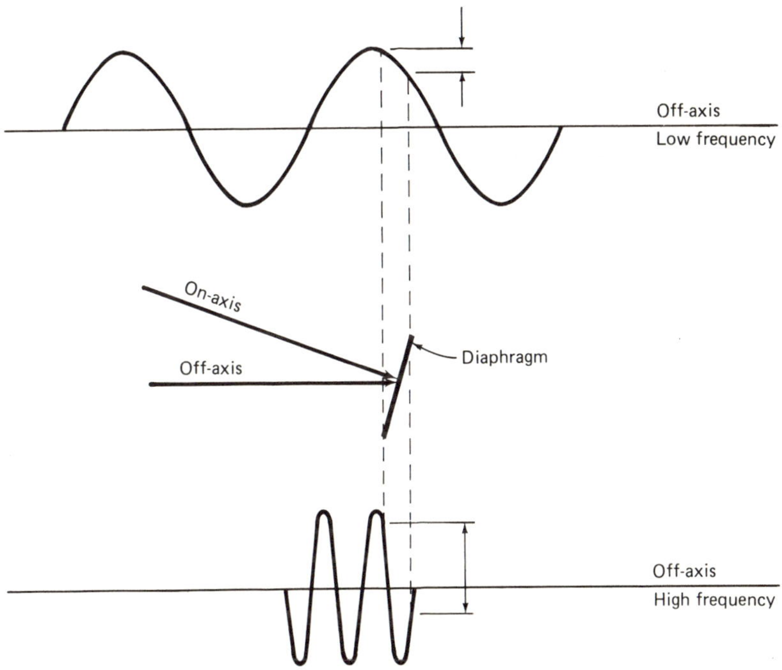

Figure 7-10. For long wavelengths (low frequencies), the angle of incidence has little or no practical effect on frequency response **(A).** However, for short wavelengths (high frequencies), off-axis incidence results in rolled off high frequency response, because different parts of the diaphragm intercept different portions of the wavelength, resulting in some cancellation of the force on the diaphragm.

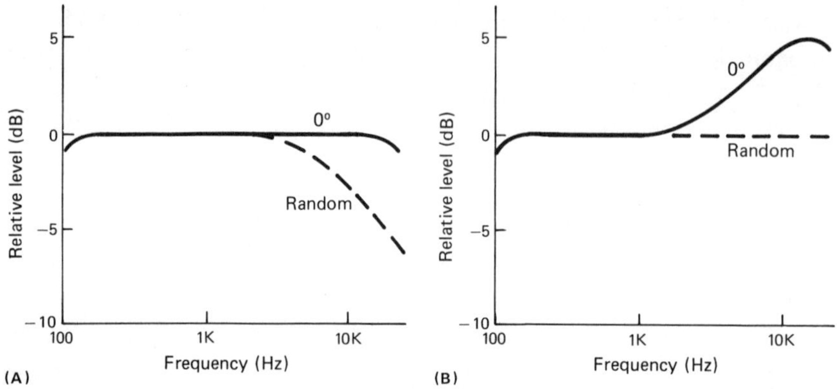

Figure 7-11. The effects of diffraction. The response of a normal microphone **(A)** is flat on axis, and rolled off for random incidence or off-axis signals. The response of a peaked microphone **(B)** displays a rising on-axis response, and a flat response for random incidence or off-axis signals.

can be useful. In an earlier era of recording, engineers appreciated the additional presence which the broad 10 kHz peak imparted to classical orchestral recording, where the microphones were required by necessity to be at some distance from the players. In that case, the broad peak acted as a kind of random incidence correction, providing a more natural overall balance. In more recent times, as instrumentation microphones have enjoyed a degree of popularity in classical orchestral pickup, some engineers have overlooked the fact that most of these microphones are flat only for zero-degree incidence, and thus will require high-frequency boost if natural balances are to be maintained at some distance from the ensemble.

DISTANCE EFFECTS

Additional high-frequency losses are incurred at distances beyond 12-15 meters (40-50 feet) due to molecular absorption by the atmosphere. Contrary to what most people have been led to believe, these losses are most significant when the relative humidity is low, as shown in Figure 7-12. As a consequence of these humidity-dependent high-frequency losses, reverberation time at high frequencies will be longer under humid conditions than under dry conditions. The effect, however, is likely to be a subtle one.

The conventional wisdom that "things sound deader on damp days" probably has more to do with the attitudes and comfort of

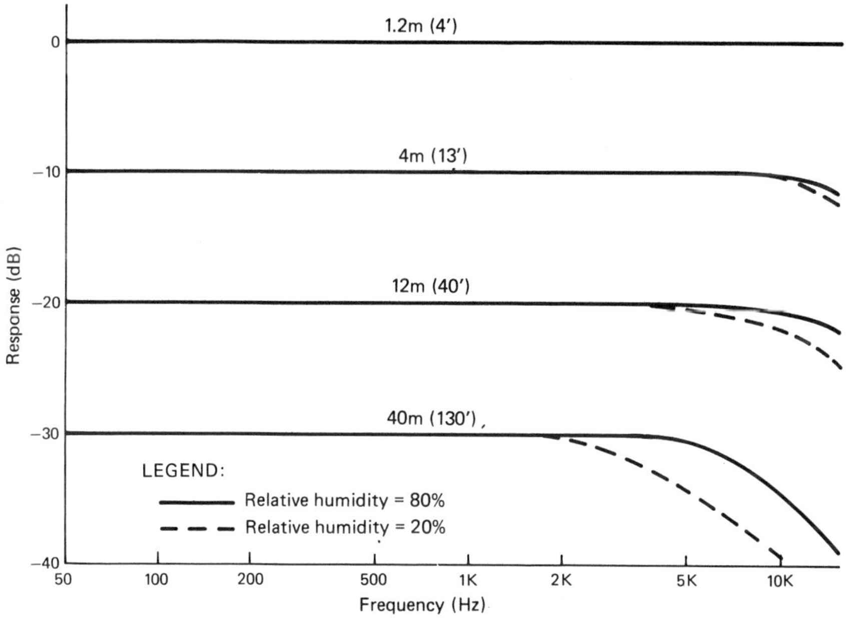

Figure 7-12. Losses due to both inverse square fall-off and atmospheric absorption at high frequencies (T = 20° C).

performers than with physics. In addition, some instruments with complex actions may feel more sluggish as a result of prolonged high humidity.

Thus, as a result of both random incidence correction and atmospheric losses, certain recording setups may require significant high-frequency boost in order to maintain proper musical balances. In particular, the recording of organs in large churches or theaters will require up to 10-12 dB boost at 10 kHz if microphones are to be used at those distances which preseve natural perspectives between the tonal resources of the instrument and the acoustical character of the hall.

CHAPTER VIII

Reflections From Nearby Surfaces

INTRODUCTION

When microphones are used adjacent to sound reflecting surfaces, there may be interference problems between the direct and reflected sounds at the microphone. The basic phenomenon is shown in Figure 8-1. Because of the path length difference, $D_2 - D_1$, there will be a time delay in the sound reaching the microphone via D_2 compared with D_1. The delay will be equal to $(D_2 - D_1)/c$, where c is the velocity of sound in air (344 m/sec or 1130 ft/sec). Thus, in the example of Figure 8-1, if $D_2 - D_1 = 1$ meter, the corresponding delay will be $1/344 = 3$ msec.

At certain frequencies, as shown in Figure 8-1B, the delay will be such that waveforms will combine at the microphone out-of-phase, with resultant cancellation. At other frequencies, as shown at Figure 8-1C, in-phase combination will result in reinfocement of the signal at the microphone.

The lowest frequency at which cancellation will occur is that frequency for which $D_2 - D_1$ is equal to one-half wavelength

$$f = 0.5c/(D_2 - D_1)$$

In the above example, f = 172 Hz.

If $f = c/(D_2 - D_1)$ then signal reinforcements will take place at integral multiples, 2f, 3f, etc. Signal cancellations will take place at frequencies of f/2, 3f/2, 5f/2, etc. (See Figure 8-2).

The degree of cancellation and reinforcement observed when signals are combined with a time offset between them depends upon their relative amplitudes, as shown in Figure 8-3. The effect is greatest when both signals are of the same amplitude, as shown at A. At B and C, we observe the effects of increasing differences in amplitude. When the amplitude ratio exceeds 4-to-1, the ratio of reinforcement to cancellation will be 5-to-3, representing a variation of 4.4 dB. For

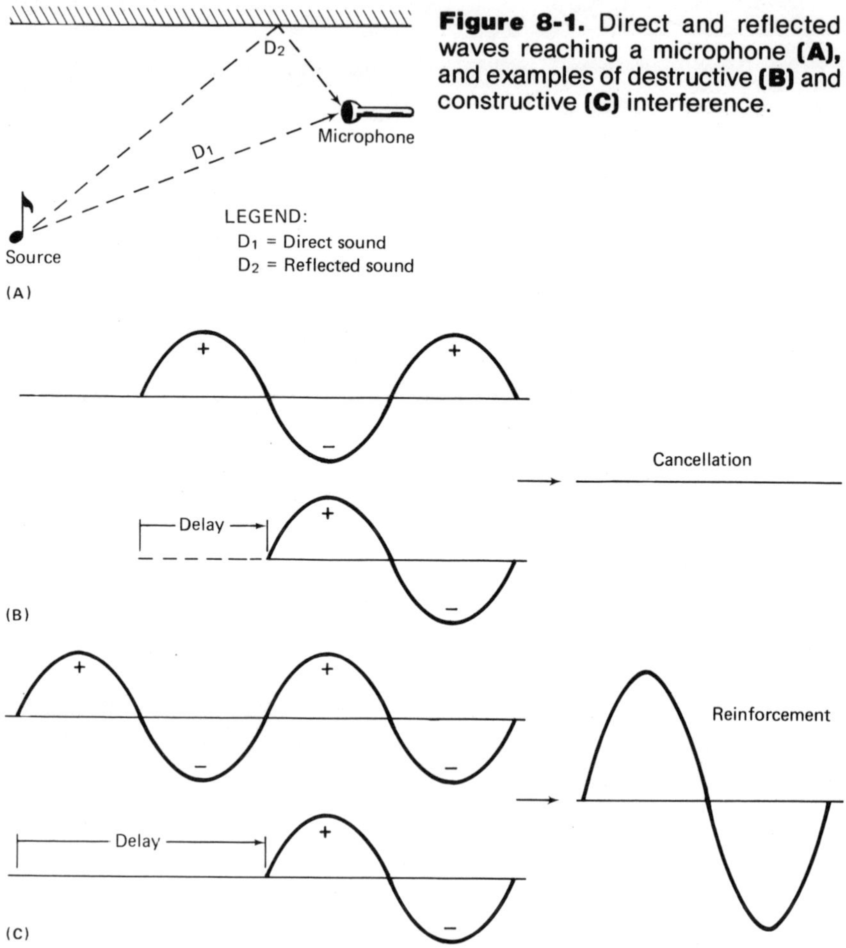

Figure 8-1. Direct and reflected waves reaching a microphone **(A)**, and examples of destructive **(B)** and constructive **(C)** interference.

any amplitude ratio, N-to-1, the difference between the peaks and dips in response will be (N + 1)/(N − 1).

The term *comb filter* is generically applied to any process in which a signal is recombined with itself through a delayed path. As we have seen, the result of this recombination is a response whose peaks and dips resemble the tines of a comb. There are three equations which define the parameters of a comb filter:

$$\text{Frequencies of reinforcement} = \frac{nc}{\text{delay}}, \text{ where } n = 1,2,3,\ldots$$

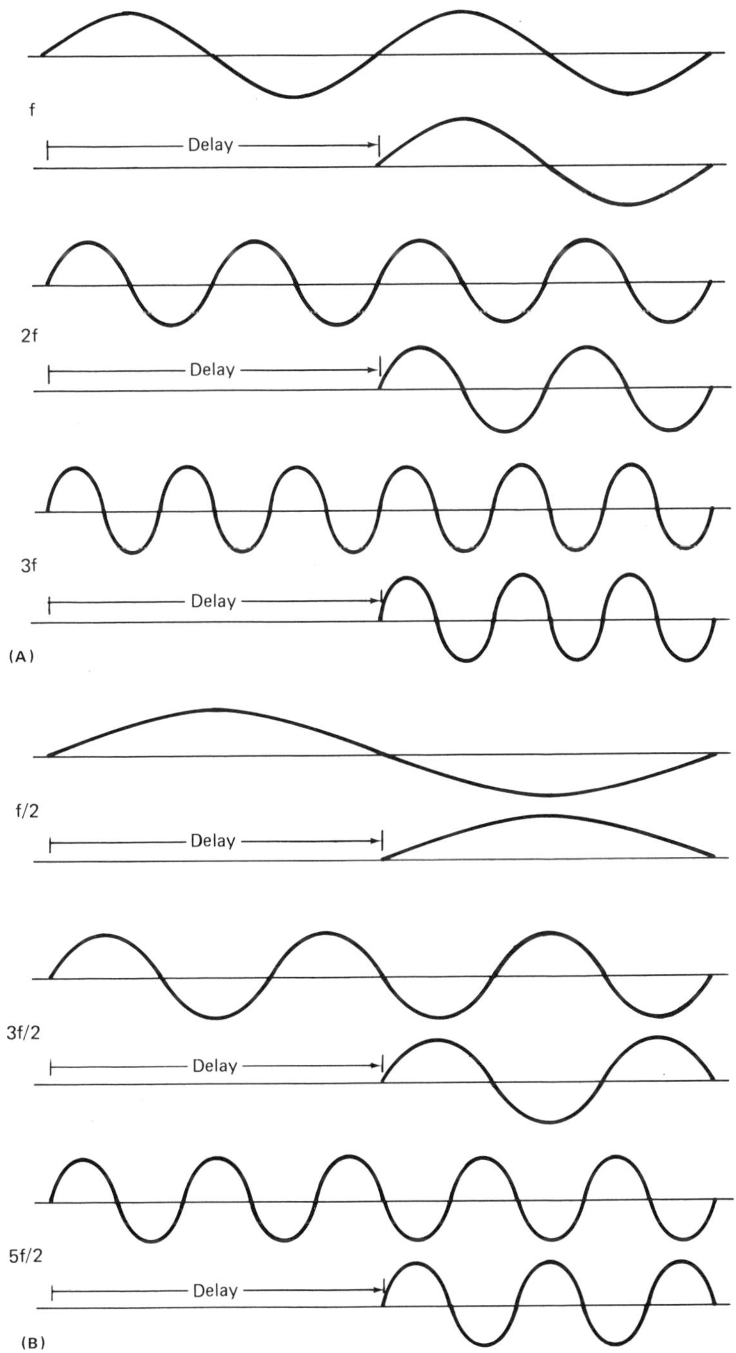

Figure 8-2. Reinforcements **(A)** and cancellations **(B)** at integral frequency multiples for a fixed delay.

$$\text{Frequencies of cancellation} = \frac{(1+2n)c}{2(\text{delay})}, \text{ where } n = 0,1,2,3,\ldots$$

$$\text{Ratio of peaks to dips in response (dB)} = 20 \log \left(\frac{N+1}{N-1}\right).$$

AUDIBILITY OF COMB FILTER EFFECTS

In almost any acoustical environment, there are significant interference effects in our perception of musical sources due to reflections. Normally, the faculty of *binaural* hearing enables us to distinguish between direct and reflected sounds. Binaural hearing, the use of both ears, allows us to determine with considerable accuracy the direction of sound sources in normal environments. It is analogous to binocular vision, the faculty of both our eyes working together to produce depth perception. A microphone, of course, receives sound at only one position, and its output signal is normally processed as a single

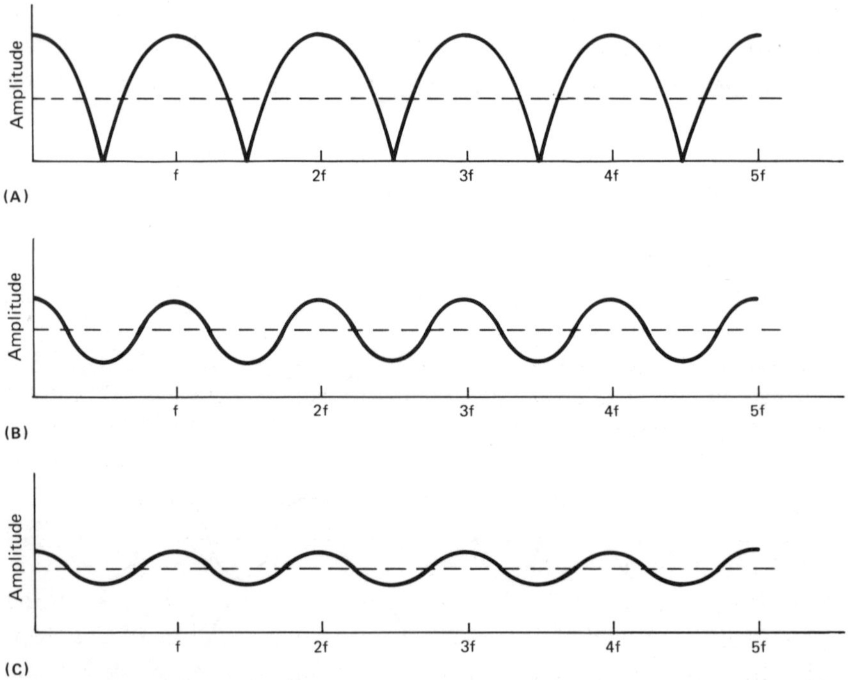

Figure 8-3. The peak-to-dip ratio of a comb filter with; both signals at the same level **(A)**; with signal level ratios of 2-to-1 **(B)** and 4-to-1 **(C)**.

channel. (Exceptions to this are special stereo and binaural microphone pairs which we will discuss in Chapter 10.) The interested engineer can demonstrate for himself the role binaural hearing plays in sorting out complex reflections:

While a performer is playing in a studio, walk up to a wall with eyes closed. There will be no doubt as to the direction of the sound source. Now, repeat the experiment with one ear blocked. Note that reflections and echoes are more prominent than before and that disturbing colorations in the sound are evident. Without binaural hearing, the brain can no longer separate the time-related multiple signals reaching the ear.

A single microphone works very much the way a single ear does, and care must be taken to ensure that it is properly placed to control the effects of reflections on frequency response.

TYPICAL PROBLEMS

Most problems with reflections arise out of improper microphone mounting for speech reinforcement applications. In Figure 8-4A, we see the usual mounting of a microphone on a podium, as well as the response due to the combination of direct and reflected sound arriving at the microphone. The solution is to mount the microphone as shown in Figure 8-4B; its smoother response is a function of the direct sound only, since reflected sounds do not reach the microphone at all.

Where microphones are permanently mounted, the above solution is easy to implement. But where microphones must be set up hastily, available hardware may not allow for flush mounting. Field broadcast applications often call for many microphones, and this will compound the problem. Where possible, *fewer* microphones should be *shared.* They should be placed as closely to the base of the podium as possible and clear of any overhanging parts of the podium.

The use of a microphone on a typical stand will place it 1-2 meters (3-6 feet) above the floor. The effect of floor reflections will vary depending on how far the speaker is from the microphone. With the speaker located at position 1 in Figure 8-5A, the effect of the floor reflection will be minimal. Because the distance ratio is so great, the delayed signal will be much lower in level, and the resulting comb filter response will exhibit a relatively small peak-to-dip ratio. For

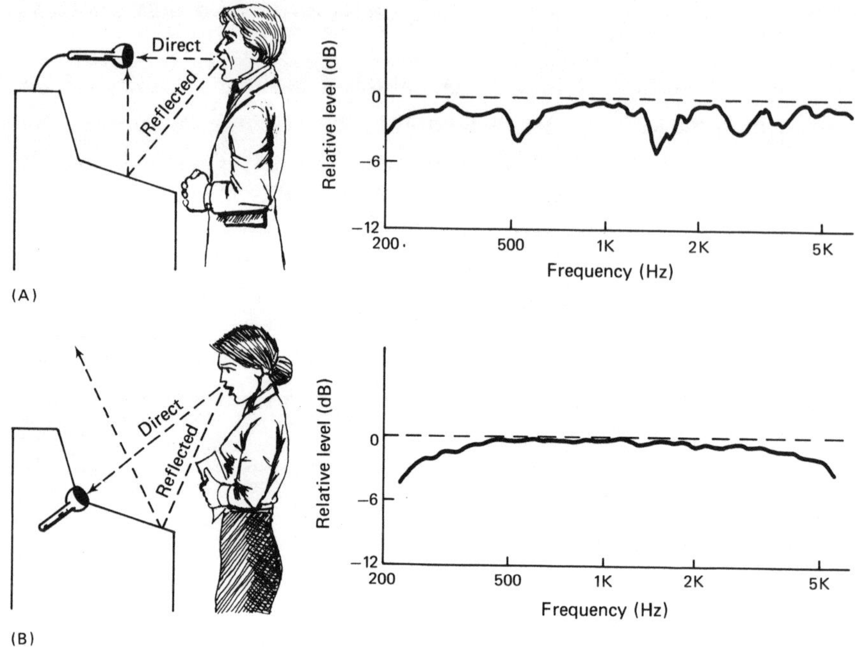

Figure 8-4. Podium microphones interference problems. Direct and reflected waves **(A)** and the resultant response for a typical podium installation. A flush-mounted microphone **(B)** eliminates interference from reflected waves, resulting in a smoother response.

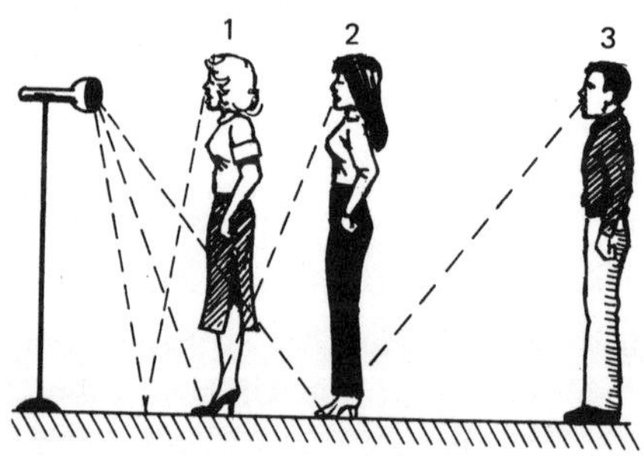

Figure 8-5. Floor stand microphone interference problems. As the working distance increases **(A)**, the frequency response deteriorates **(B)**, due to comb filter effects.

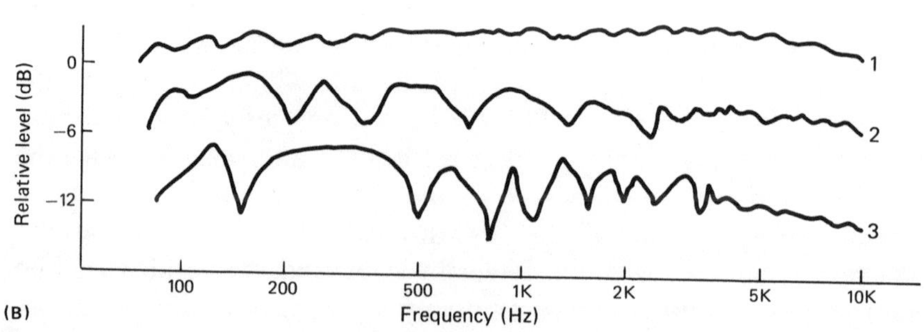

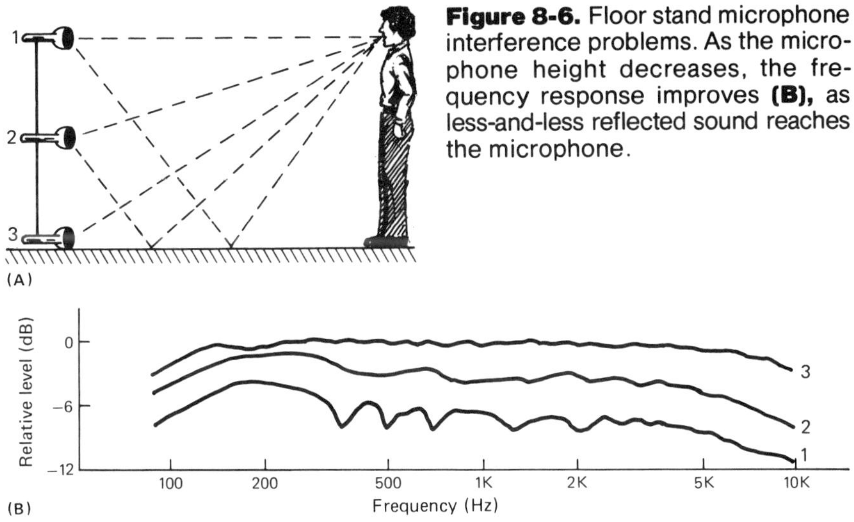

Figure 8-6. Floor stand microphone interference problems. As the microphone height decreases, the frequency response improves **(B),** as less-and-less reflected sound reaches the microphone.

speakers located at positions 2 and 3, the response becomes increasingly worse, since the distance ratio becomes progressively smaller.

Obviously, for more-distant speakers we must find some way of minimizing the comb filter effect. At Figure 8-6 we show the effect of progressively lowering the microphone for a distant source so that it is effectively flush-mounted at the floor. Under this last condition, the delay path is practically nonexistent and smooth response results.

Again, this solution is easy to implement for fixed applications such as stage presentations, where interfering audience noises will be some distance away. In closer quarters it will not be satisfactory, and we will have to find better solutions to the problem.

Similar problems occur in houses of worship where speech at the altar is to be reinforced. The flush-mounted microphone will always ensure good results, as shown in Figure 8-7.

The traditional approach to speech pickup of conferences with panelists around a table calls for the ubiquitous microphone stand, as shown in Figure 8-8. While this accessory is very useful when the microhone must be movable, it is passé for fixed microphone placement on a table or desk. It was perhaps a necessity when microphones were bulky devices; today, microphones are quite small, and there are numerous flush-mounting methods available.

It is important to note that only omni-directional microphones should be used for flush-mounted applications. Directional microphones will give rise to aberrant response when used in these applications

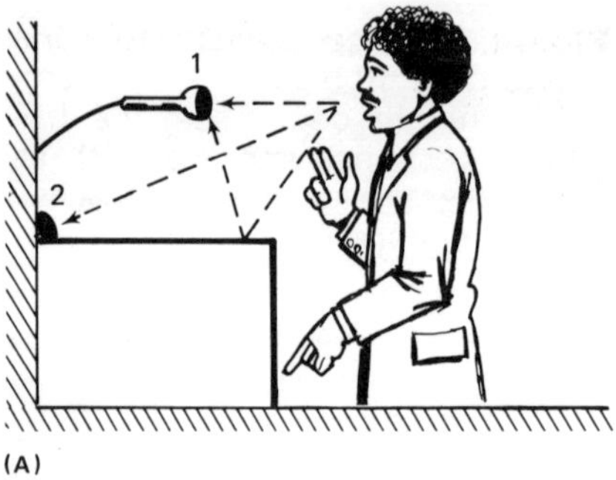

Figure 8-7. Altar microphone interference problems. As in Figure 8-6, the microphone's position **(A)** affects the frequency response **(B)**.

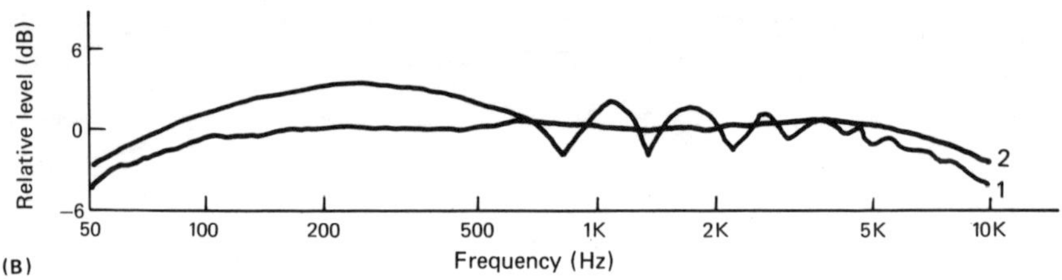

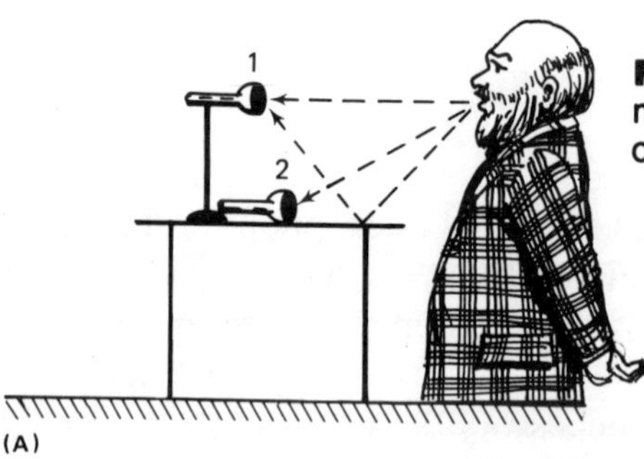

Figure 8-8. Desk microphone placements **(A)**, with resultant response curves **(B)**.

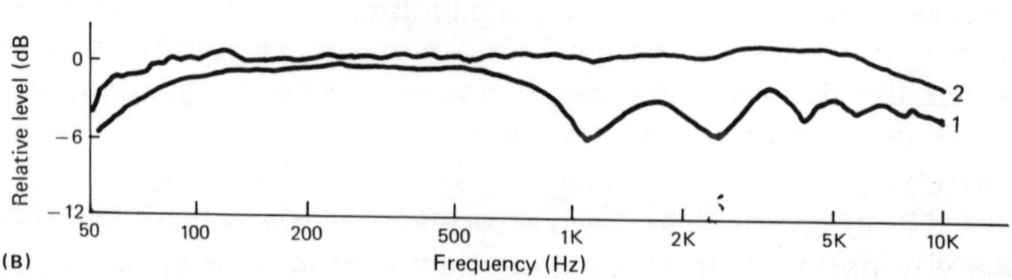

due to the various back openings to the diaphragm. Flush mounting will interfere with these openings and alter the response of the microphone, often to a profound degree.

PERSONAL MICROPHONES

Where the application allows, personal microphones overcome all of the comb filter effects we have discussed in this chapter. A personal microphone is one worn by the user. There are three main types as shown in Figure 8-9: the lavalier or lanyard type, the lapel or tie-tack type, and the boom type. Most lavalier microphones are relatively bulky and often give rise to noises due to their brushing against clothing.

The boom type is a specialized deviced used in fairly active and noisy environments. Electret lapel microphones have become very attractive as the electret capacitor system has become both smaller and less expensive.

For realistic speech reproduction, lavalier and tie-tack microphones must be internally equalized in a specific way. Because they are located close to the chest, their low-frequency response must be rolled off to bring chest cavity response in line with normally perceived speech. Since they are located quite far off the normal axis of the voice, their high-frequency response must be boosted accordingly. The result is the ideal response curve shown in Figure 8-10, which results in extremely natural voice quality.

THE WIRELESS MICROPHONE

For those appications where the performer or speaker must move around with complete freedom, a wireless microphone may be essential. These microphones feed directly into a small portable RF modulator. The transmitting antenna is often no more than a short piece of wire hidden in the performer's clothing. FM transmissions are received at some off-stage position, and thc signal is amplified and routed as usual. Details are shown in Figure 8-11.

While the best wireless microphones are noise- and trouble-free, many less-expensive and earlier models are subject to RF interference, especially in urban areas. Their use is best specified only after careful surveys have been made.

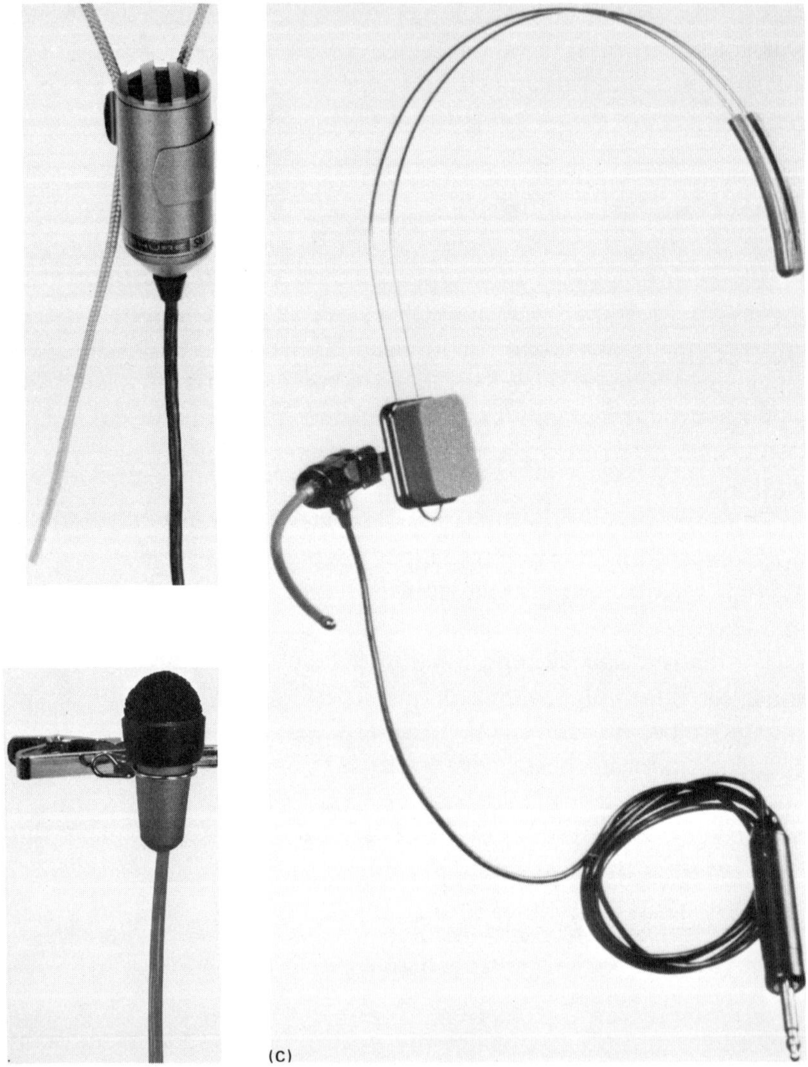

(A) (B) (C)

Figure 8-9. Personal microphones. **(A)** lavalier style *(Shure Bros photo);* tie-tack or lapel style **(B);** boom style **(C)** *(Electro-Voice photos).*

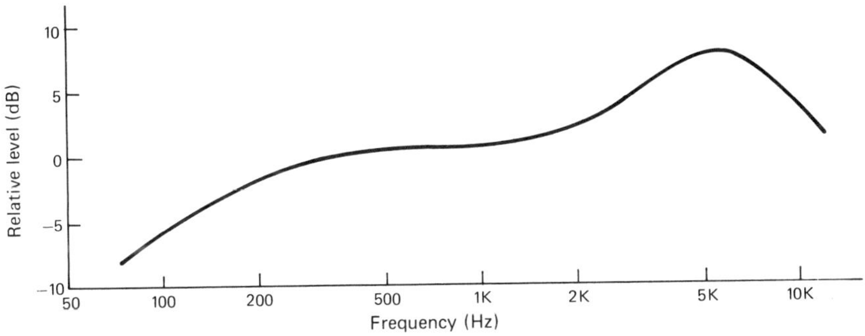

Figure 8-10. The frequency response of a lapel-style microphone showing typical low-frequency roll-off and high-frequency peak.

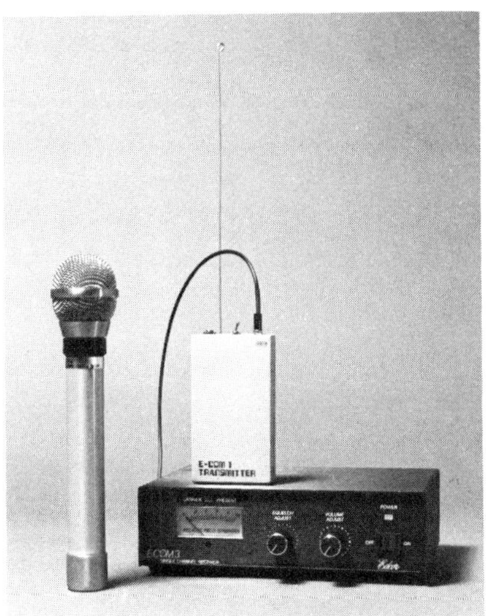

(A)

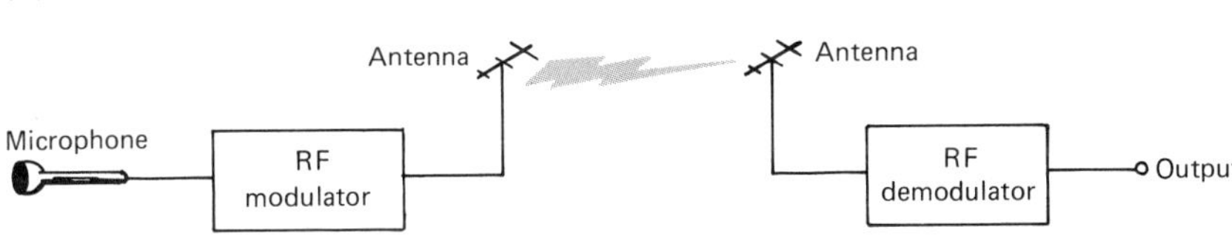

(B)

Figure 8-11. A typical wireless microphone **(A)** and block diagram **(B)** showing modulator and demodulator *(Edcor photo).*

CHAPTER IX

Multi-Microphone Interference Problems

INTRODUCTION

The inexperienced engineer normally has a tendency to use too many microphones, and it is out of this tendency that most problems of microphone interference arise. The mechanism of microphone interference has much in common with interference phenomena due to reflections discussed in the preceding chapter. Figure 9-1 outlines the problem.

When two microphones are to be fed into a common mixing bus, any difference in their distances from a common sound source will result in comb-filter response of that source. The equations developed in the previous chapter apply directly to the problem. Assuming values of $D_1 = 1$ m and $D_2 = 2$ m, note that the result will be a comb-filter response with a peak-to-dip ratio of 9.5 dB. In this graph, we are showing the theoretical comb-filter response; in actual practice the peaks and dips would not be as pronounced, especially at higher frequencies.

Note that the frequency interval between successive peaks and dips remains a constant 172 Hz. When plotted on a *log-frequency* basis, as in Figure 9-1, the peaks and dips cluster together with rising frequency, producing a very dense array of peaks and dips in the region above 2 kHz.

Those readers who are of an experimental turn of mind are urged to set up the microphone array shown in Figure 9-2. While a speaker moves back and forth from position 1 to position 5, listen carefully to the sound of both microphones combined. Only when the speaker is very close to either microphone—or *precisely* midway between them—will the sound be natural. Note that the peak-to-dip ratio is greatest at the two positions, 2 and 4, adjacent to the center position.

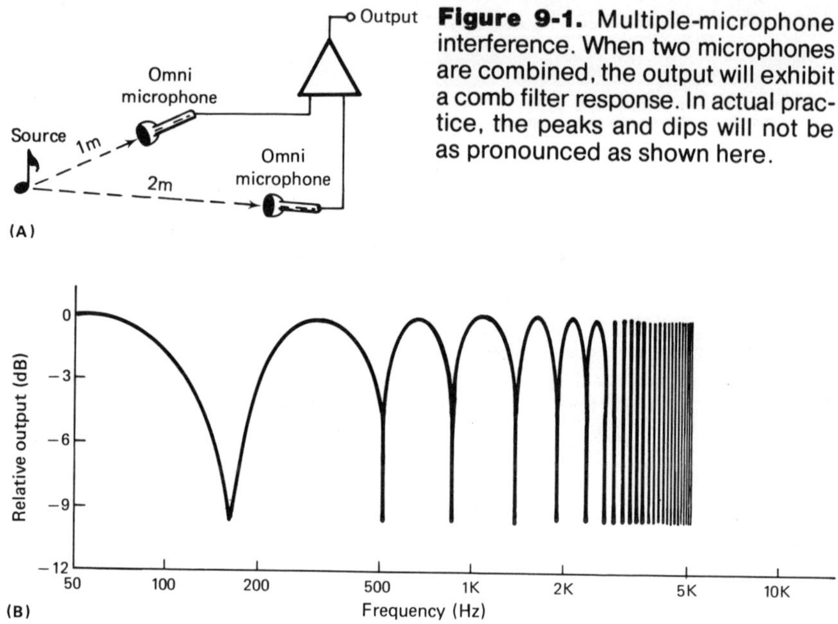

Figure 9-1. Multiple-microphone interference. When two microphones are combined, the output will exhibit a comb filter response. In actual practice, the peaks and dips will not be as pronounced as shown here.

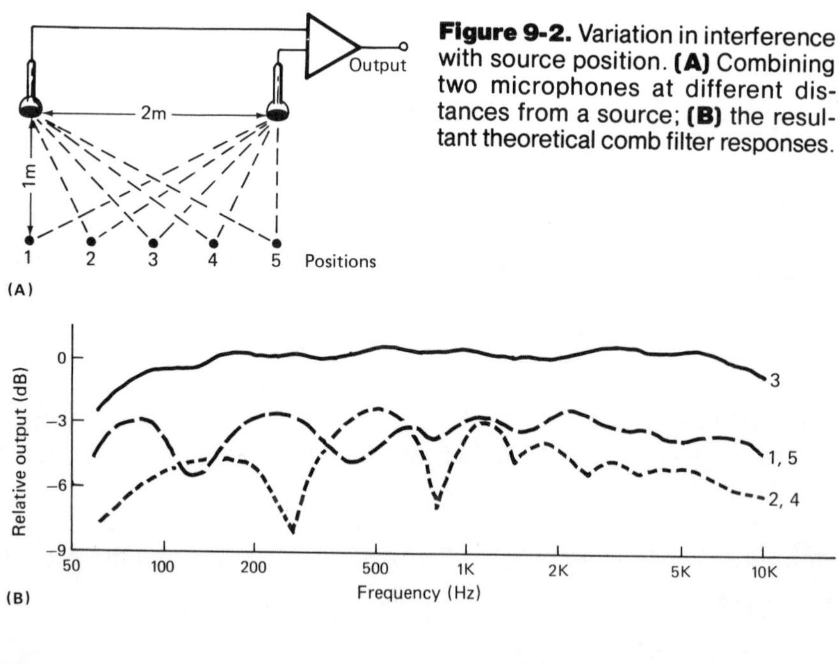

Figure 9-2. Variation in interference with source position. **(A)** Combining two microphones at different distances from a source; **(B)** the resultant theoretical comb filter responses.

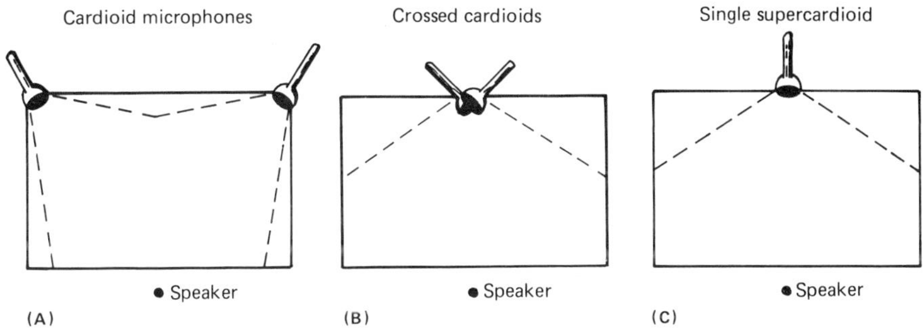

Figure 9-3. The use of podium microphones. **(A)** improperly placed cardioid microphones; **(B)** a properly-placed cardioid pair; **(C)** a properly-placed single microphone.

PODIUM PROBLEMS

We have all seen podium microphones arrayed as shown in Figure 9-3A. There is *never* any justification for this. As we have noted, only when the speaker is exactly in the middle—which he rarely is—will the response be smooth. In a far-better array, two crossed cardioids will not interfere, but will simply exhibit a broadened acceptance angle (Figure 9-3B).

But why use two microphones at all? The argument in favor of reliability due to redundancy is a weak one inasmuch as microphones are among the most reliable elements in any sound system. The best solution is to use a single super-cardioid. Here (Figure 9-3C), the acceptance angle is well in excess of 120 degrees, and the speaker is ensured of adequate pick-up, whatever his position behind the podium.

CHOOSING A WORKABLE DISTANCE RATIO

The distance ratio of two microphones picking up the same sound will determine the peak-to-dip ratio in the resulting comb-filter response. In Figure 9-4, we show this ratio in decibels for various combinations. Note that beyond a 3-to-1 ratio, the peak-to-dip ratio decreases rather slowly.

Accepted practice in the recording studio is that a ratio of at least 3-to-1 suffices to minimize interference effects, while allowing the performers to play within reasonable distances of each other. At a 3-to-1 distance ratio, the resulting comb-filter response with a 6-dB peak-to-dip ratio may seem extreme. Actually, it is not, and the use of directional microphones can only improve the situation.

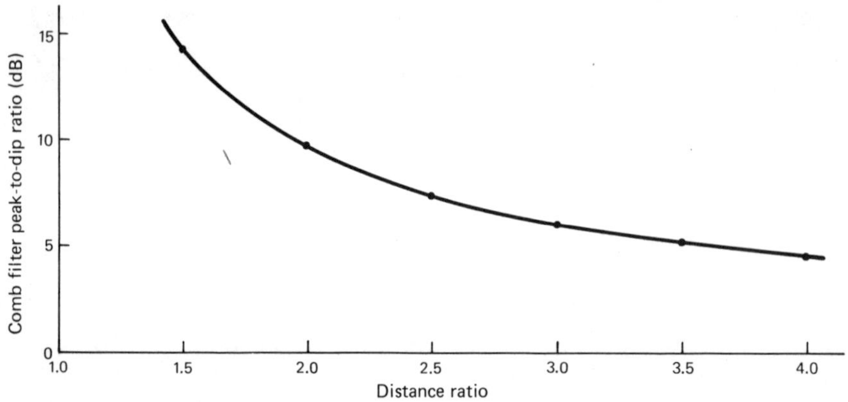

Figure 9-4. The effect of distance ratio on comb filter response. As the distance ratio increases, the comb filter effect diminishes.

STUDIO PROBLEMS

No matter how many microphones are used, the entire ensemble is ultimately reduced down to a stereo pair of tracks. There may also be a requirement for monophonic compatibility when the stereo pair is combined. Thus, it is a certainty that cancellations will take place if special pains are not taken to prevent them. The most straightforward case is when two musicians are playing instruments of similar output, as shown in Figure 9-5A. If both microphones are omni-directional, then application of the 3-to-1 rule will alleviate the problem.

However, if one instrument is significantly lower in level than the other (Figure 9-5B), the problem will be worse; the raised level required for the microphone picking up the softer instrument will result in an even more-violent comb-filter response. Of course, it would be bad practice to place such disparate instruments so close to each other when the intention is to pick them both up with good separation. A combination of wider spacing and directional patterns will be required to correct the problem, as shown in Figure 9-5C.

A strict application of the 3-to-1 rule in the studio will need to be tempered by many other considerations. Within a sectional group—or a group of instruments of similar output level—the 3-to-1 ratio may well apply. Where there will be considerable gain manipulation of individual tracks for whatever reason, then a greater ratio, perhaps 5-to-1, will be needed in order to minimize leakage from loud instru-

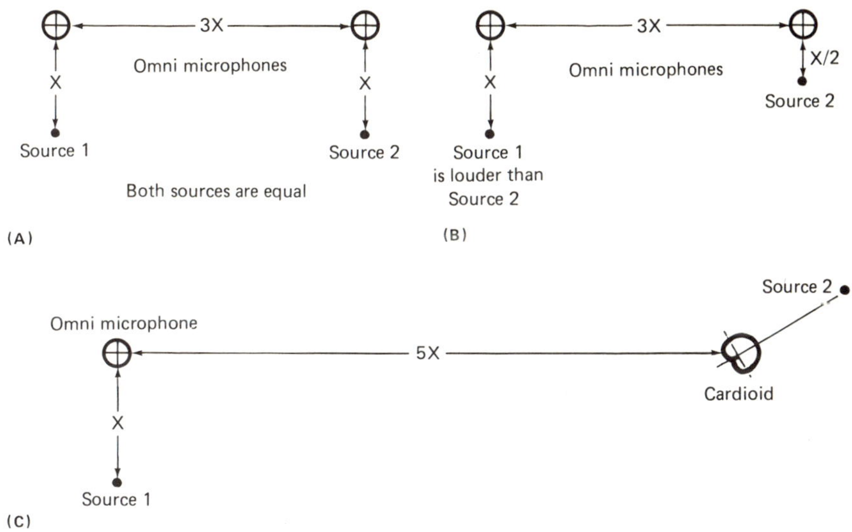

Figure 9-5. The 3-to-1 ratio in the studio, for sources of equal **(A)** and unequal **(B)** acoustical level. **(C)** The application of a greater distance ratio (5-to-1) and a directional polar response for improved isolation with two sources of widely differing acoustical level.

ments into these microphones whose gains may have been momentarily increased.

Figure 9-6 gives a layout of a hypothetical studio setup. Note that there are several scales of the 3-to-1 ratio. Not only will softer instruments require closer microphone placement, they will have to be spaced at whatever distance is required in order to avoid excessive leakage from the louder instruments into the microphones assigned to the softer ones. The closer placement of microphones to the softer instruments will be helpful in this regard, since their gain will not have to be so great as it would for more-distant placement. This invokes a host of other problems, including proximity effects and fluctuations in level due to normal movement of the players at such close distance to the microphones.

CONCLUSION

The reader will learn in the following chapter that overlap of microphone pickups, and its consequent interference effects, is not all bad. Many of the effects of spatial perspective implicit in any good stereo-

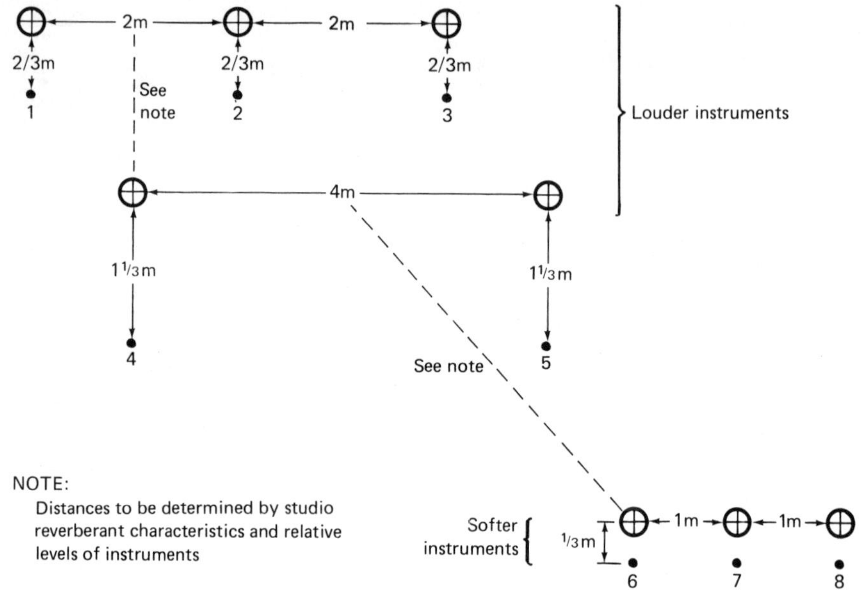

Figure 9-6. The 3-to-1 ratio in the studio, for multiple sources of varying acoustical levels.

phonic recording depend upon an ensemble of what are often referred to as "leakage" signals, those which are picked up via secondary microphone channels. Studio experience is essential in developing a feel for the proper balance between signal isolation for the sake of clarity, and signal overlap between microphones for the sake of stereophonic integrity.

CHAPTER X

Stereo Microphone Techniques

INTRODUCTION AND HISTORY

In this chapter we begin our discussion of how ensembles of microphones are able to convey information relating to the geometry and spatial perspectives of the environments in which they are used.

The earliest experiments in spatial sound reproduction date from the late 1800's, when telephone receivers and earphones were used in conveying a sense of auditory perspective through the use of an *artificial head.* The technique is shown in Figure 10-1. Here, microphones are placed inside the artificial head at ear position. At some remote location, a listener wears headphones, each corresponding to one of the microphones. In this method of sound transmission, referred to as *binaural sound,* the ears of the listener are effectively transported to the performance environment itself. Despite the shortcomings of those earliest transducers, it is reported that this system conveyed a remarkable sense of realism.

In a later section of this chapter we will discuss binaural sound and further extensions of it in more detail.

COINCIDENT MICROPHONE TECHNIQUE

Stereophonic sound refers to the pick-up of sound with an array of at least two microphones and subsequent replay over a pair of loudspeakers in such a manner as to convey some sense of the original left-right, fore-aft spatial relationships of the original performance. The earliest experiments in stereo were probably carried out by Alan Dower Blumlein in England in the early nineteen thirties. Those experiments made use of what has become generally known as the *coincident* microphone technique, an approach where the microphones in the pick-up array are clustered closely together.

Just a few years later, Bell Telephone Laboratories in the United

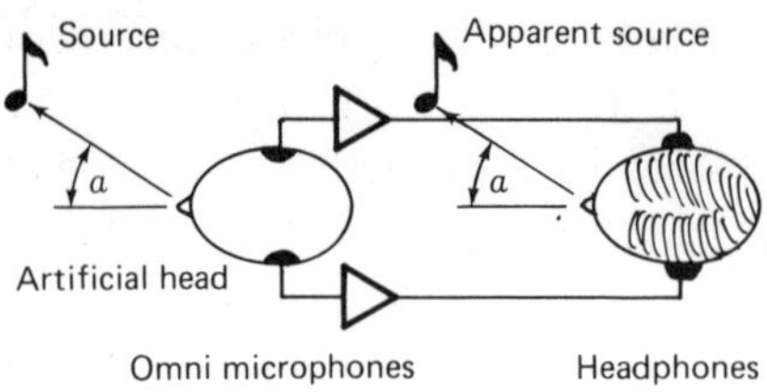

(A)

Figure 10-1. The principle of binaural sound. **(A)** By means of microphones placed in an artificial head, directional cues may be transmitted to a listener wearing headphones. **(B)** A typical artificial head setup *(Sennheiser photo).*

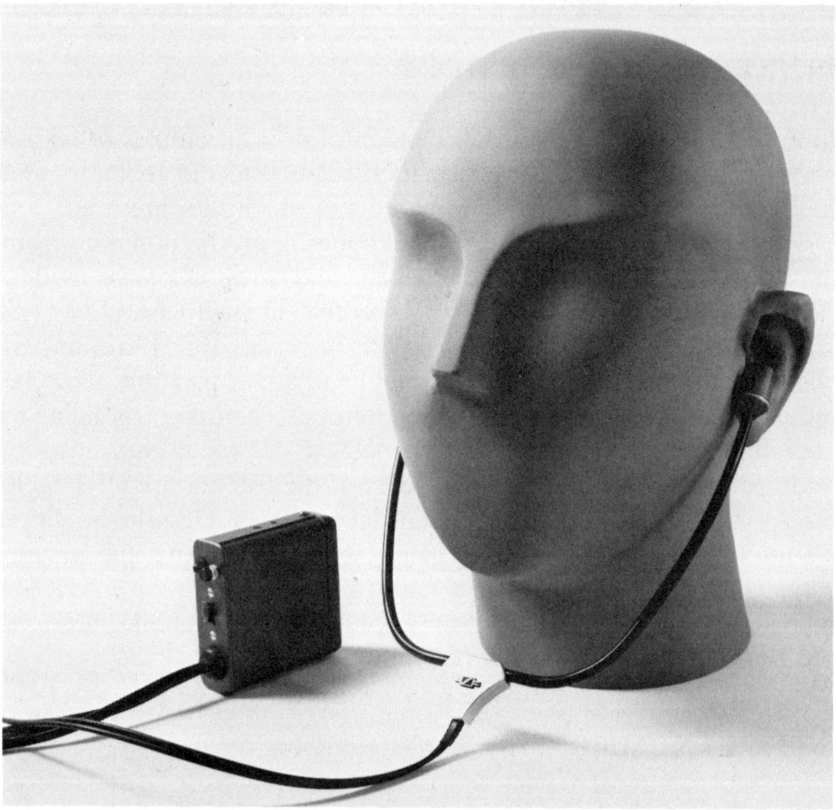

(B)

States embarked upon an ambitious program which incorporated auditory perspective and multi-channel sound transmission, developed along with significant improvements in amplifiers, loudspeakers and microphones.

The art remained fairly constant for the next 20 years. Only during the post-World War II period did the magnetic tape recorder provide the impetus for the record industry to experiment with stereo. Recorded tapes were introduced in the early fifties, and the stereo disc was introduced in the late fifties. That decade saw a rapid expansion of both the technical and aesthetic bases of stereo recording, and the art was extended beyond its role of merely documenting acoustical events, into one of creating new and original sounds in the studio.

In recent years, the introduction of 16- and 24-track recorders has broadened that base further. But too often, the simple lessons of Blumlein have been forgotten, and so it is with a review of his work that we begin our study.

BLUMLEIN'S TECHNIQUES

One of Blumlein's basic stereophonic microphone arrays is shown in Figure 10-2. A pair of crossed bi-directional microphones is oriented

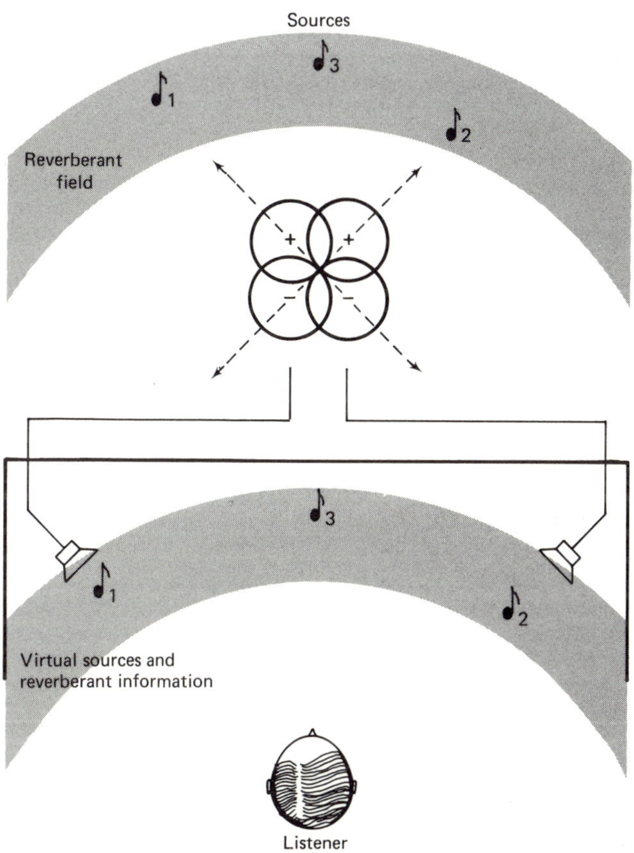

Figure 10-2. Blumlein's stereophonic technique. Crossed figure-8 microphones respond to sound arriving from all directions.

so that the axis common to the two positive lobes is aimed toward the center of an ensemble of performers. Because of the precise nature of the cosine polar response of the two bi-directional microphones, a sound source at position 1 will be picked up only by the left-oriented microphone. A similar relationship occurs with the right microphone and a sound source at position 2. For a sound source at position 3, there will be equal response at the two microphones, and the sound source will appear in the playback loudspeaker array to be directly in the center. Reverberant energy, since it comes from all directions, will be reproduced by the pair of microphones as coming from the space between the loudspeakers quite evenly.

In this way, a pair of bi-directional microphones crossed at an angle of 90 degrees behaves exactly like a *panoramic potentiometer* (panpot). The panpot is a signal-processing device which routes a single input into two outputs so that the apparent position between loudspeakers becomes a function of the physical position of the panpot itself. The gradual cross-fading of the input between one output and the other provides a smooth transition of the reproduced image between the two loudspeakers in the playback array. Details of a two-channel panpot are shown in Figure 10-3.

From our discussion in the last chapter, we will note that there are no interference problems when the outputs of both microphones are mixed. Since they are coincident—occupying virtually the same position—their summation represents an accurate monophonic signal.

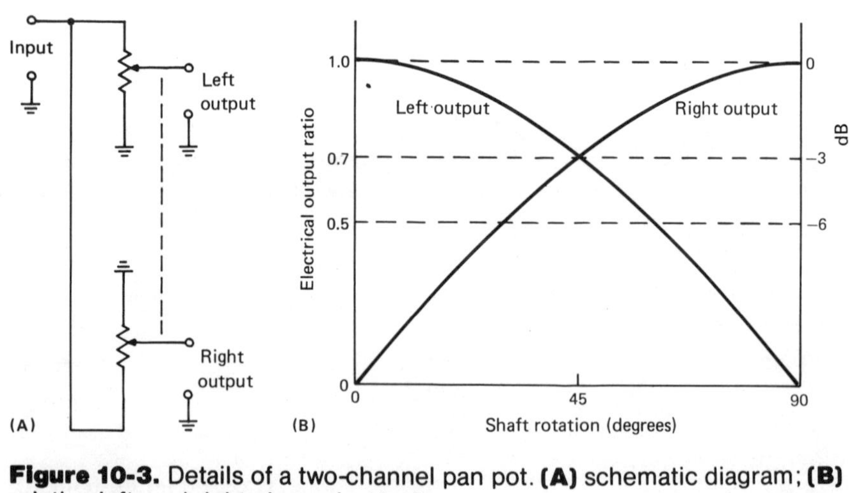

Figure 10-3. Details of a two-channel pan pot. **(A)** schematic diagram; **(B)** relative left- and right-channel outputs.

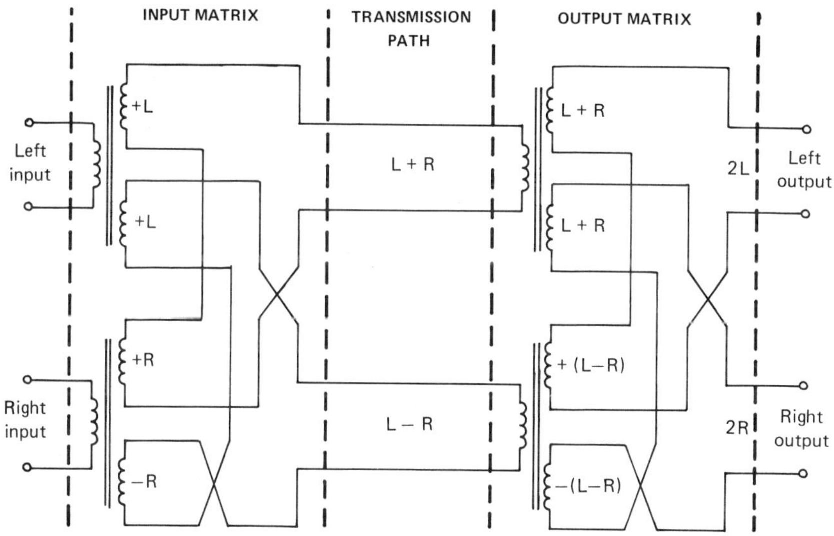

Figure 10-4. A sum-and-difference matrixing system. For certain applications such as FM stereo broadcasting, it is convenient to transmit stereo as a *sum* channel (which is used for monophonic transmission) and a *difference* channel. The two can be recombined as required for normal stereo reproduction.

This property of the crossed Blumlein-pair leads us to the concept of *sum* and *difference* transmission of the two signals, as shown in Figure 10-4. Through the use of what are called *matrixing* transformers, the left and right microphone outputs are added and subtracted to derive sum and difference signals. An additional pair of transformers undoes the process, and the left and right signals are recovered.

The process shown in Figure 10-4 forms the basis of stereophonic FM broadcasting, wherein a mono receiver picks up only the sum signal. The same condition exists in the playback of a stereophonic disc with a monophonic cartridge responding only to the lateral, or sum, signal.

Fortunately, those recording engineers wishing to make use of Blumlein's technique have access to so-called *stereo microphones.* These are available from a number of manufacturers, and they contain within one housing a pair of Braunmühl-Weber capacitor elements which are physically, as well as electrically, adjustable. The engineer can adjust both the precise polar response of either element, as well as the angular relationship between them.

VARIATIONS ON BLUMLEIN: THE M-S TECHNIQUE

The M-S (Middle-Side) technique uses a single forward-oriented microphone (M) as the sum signal, and a bi-directional microphone oriented at 90 degrees (S) as the difference signal. In Blumlein's early work, both microphones were bi-directional; today it is common practice to use a uni-directional (or even an omni-directional) microphone for the sum (M) signal.

When the outputs of the M-S pair are resolved by means of a sum-and-difference matrixing transformer, as shown in Figure 10-5, the result is a response equivalent to a pair of left-right oriented patterns, each of which resembles a super-cardioid pattern. Note that no input matrix is required, since the M-S pair performs this function itself.

Further elaborations of the M-S technique are *width* and *position* controls, as shown in Figure 10-5C. The width control acts by varying the amount of difference signal in the program. As the difference signal is attenuated, the reproduced program will appear to the listener to be more monophonic, or center-oriented. When the difference signal predominates, its out-of-phase components effectively broaden the signal. The effect of the position control is to shift the entire reproduced array from left to right, as required by the program requirements at hand.

Figure 10-6 shows the effective X-Y microphone patterns for various M-S ratios. The reader should remember that in the X-Y representation, the microphones are actually coincident. They have been shown separated in this figure merely for graphical clarity.

Typical usage of multiple M-S pairs of microphones is shown in Figure 10-7. Here, the main pickup pair is centered and its width adjusted to occupy the entire stereophonic playback stage. Sectional pairs would be positioned as required, and their width controls set for the appropriate effect in the overall picture.

A pair of sum-and-difference microphones resolved into left-right signals is often referred to as an X-Y pair. Figure 10-8 shows equivalent X-Y pairs forms for a variety of sum-and-difference combinations.

SUBJECTIVE CHARACTERISTICS OF COINCIDENT TECHNIQUES

The chief virtue of any coincident technique is the absolute integrity of the panning or positioning function which it provides; particular instruments seem to originate in the playback loudspeaker array at

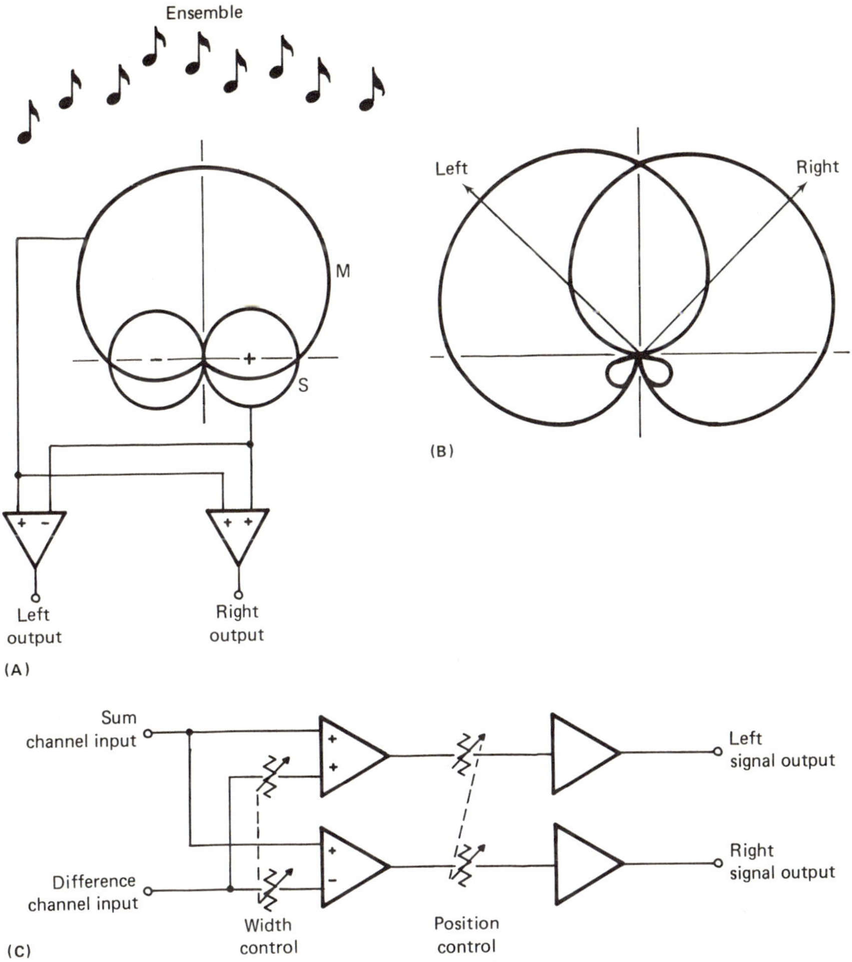

Figure 10-5. The M-S coincident microphone technique. **(A)** block diagram; **(B)** the resultant left- and right-channel outputs; **(C)** width and position controls for use with matrixing amplifiers.

their proper points. And, as we have stated, reverberant information seems to be evenly spread between the two playback loudspeakers.

Yet in spite of this high degree of spatial accuracy, coincident stereo microphone techniques are not widely used today. One reason is that the approach is highly dependent upon a precise distance relationship between the performers and their relation to the acoustical environment. The right combination is a rarity. Even when the right conditions do exist, many recording engineers and producers feel

M–S ratios where M = A + B_M cos θ (cardioid) S = B_S cos θ (figure 8)		Effective x –y microphone patterns and included angle	Description, and [polar equation o effective patterns
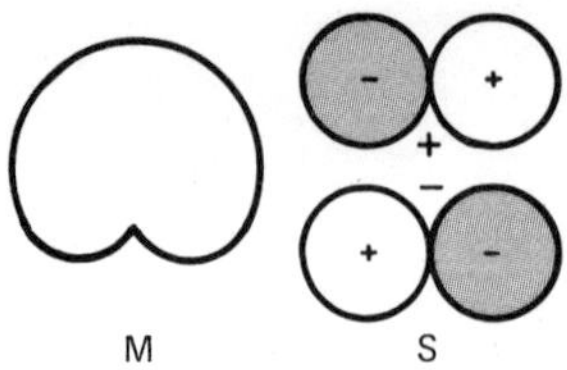			
B_M	B_S		
.5	0	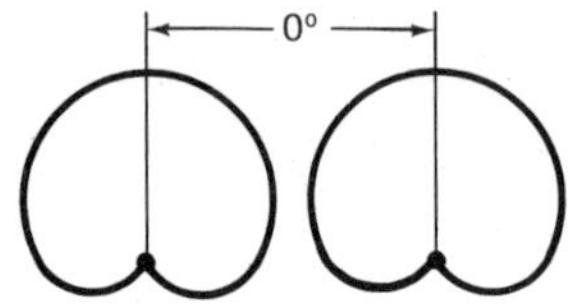	Mono signals [.5 + .5 cos θ]
.480	.200	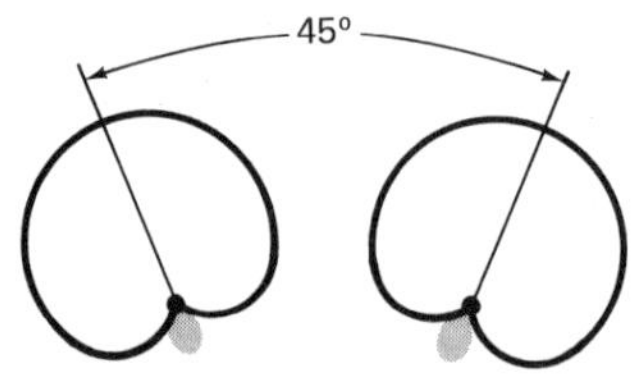	Little stereo separatic Narrow spread [.480 + .520 cos θ]
.414	.414	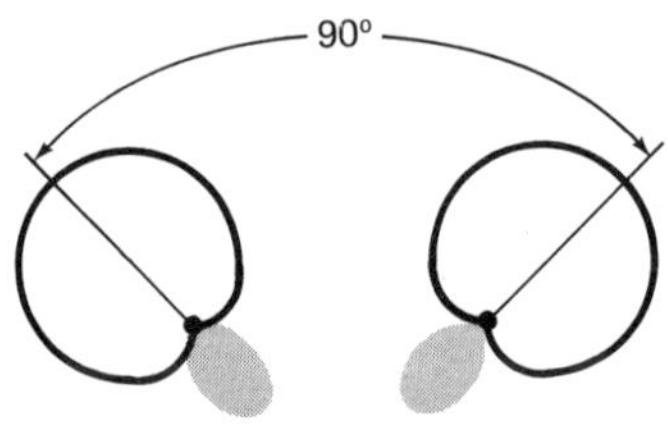	Optimum stereo sepa [.414 + .586 cos θ]
.277	.668	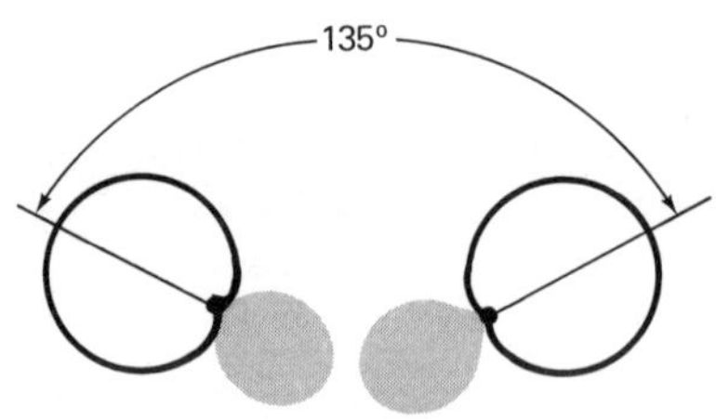	Exaggerated stereo in somewhat unstable [.277 + .723 cos θ]
0	1	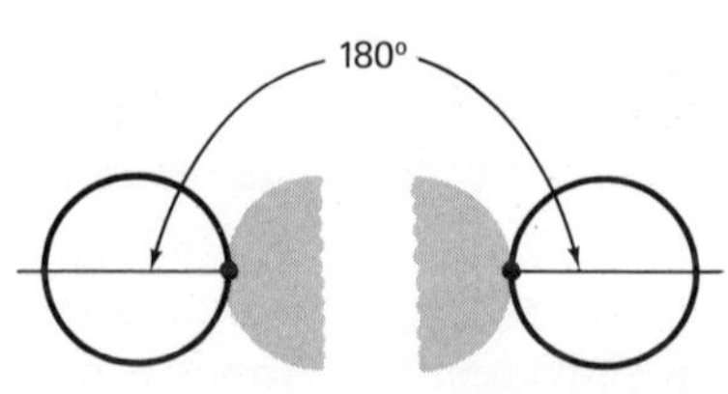	Out-of-phase mono s [1 cos θ]
Values of B_M and B_S are chosen so that the effective x–y patterns are normalized (0dB down) on-axis		Shaded area indicates polarity reversal of output	

Figure 10-6. A family of effective X-Y pairs for various M-S ratios.

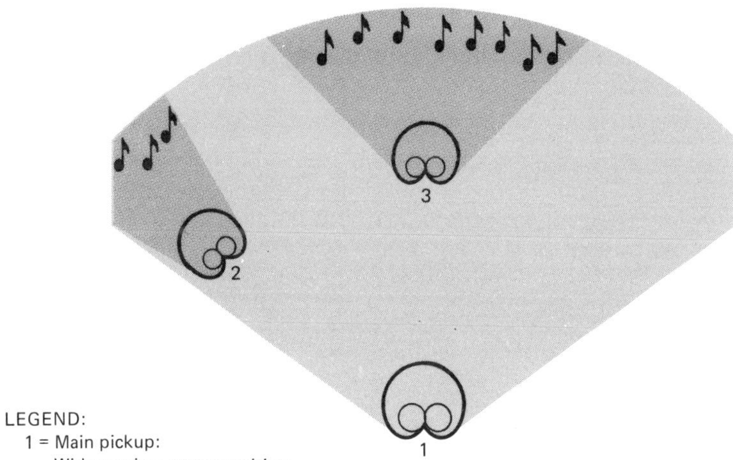

Figure 10-7. The use of width and position controls in M-S recording.

that the coincident technique lacks the degree of warmth and depth which even a slight spacing apart of the microphones provides.

SPACED-APART TECHNIQUES

Experiments by the French National Broadcasting Organization (ORTF—*Office de Radiodiffusion-Television Française*) have led to the X-Y microphone array shown in Figure 10-9. This technique may be best described as *quasi-coincident;* at low frequencies (long wave lengths), the two microphones effectively occupy the same location. At frequencies above 500 Hz, the microphone array responds to time delays due to the finite spacing of the pair; the listener senses not only these time-delay cues, but also the intensity difference, which is primarily a function of each microphone's polar response. Most listeners sense this as a feeling of space and "airiness" in the reproduced sound. A similar technique has been adopted by the Dutch Broadcasting Organization (NOS—*Nederlandsche Omroep Stichting),* making use of a pair of X-Y cardioids spaced 30 cm apart at an angle of 90 degrees.

A similar technique was proposed some years ago by Madsen. He located a pair of bi-directional ribbon microphones as shown in

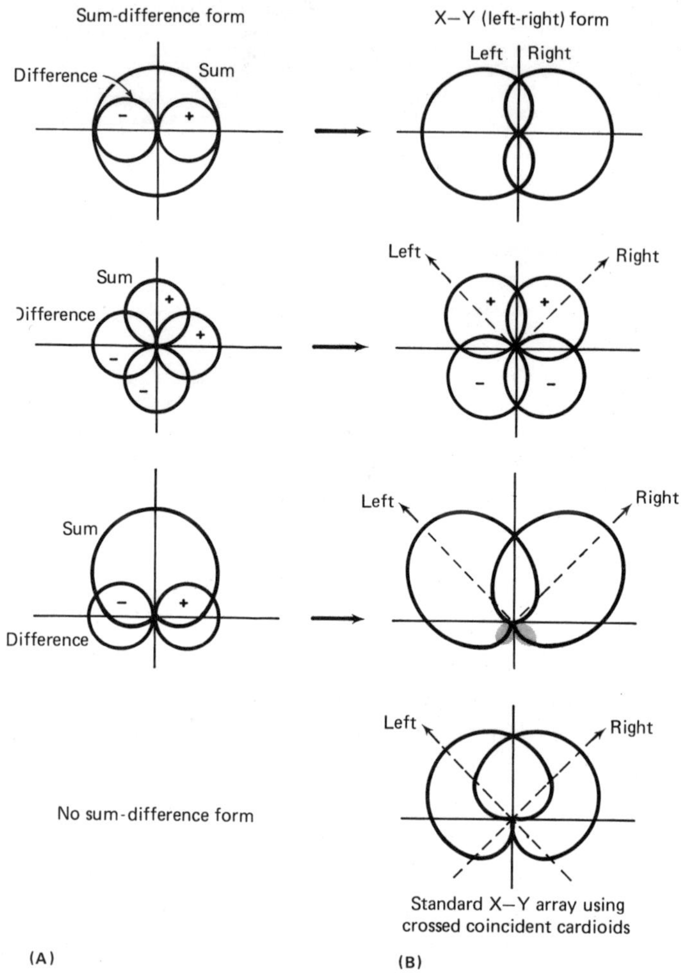

Figure 10-8. Various sum-difference/X-Y pairs. The sum-and-difference outputs from a pair of microphones **(A)** resemble the X-Y pairs seen at **(B)**.

Figure 10-10. A baffle placed between the pair modified their response such that the patterns became more directional at high frequencies.

As we have observed in the discussion of coincident microphones, a pair of directional microphones can pick up stereo information based upon intensity relationships alone. Of course, a coincident pair of omni-directional microphones would pick up no stereo information; the outputs of both microphones would be virtually indentical. In order to work satisfactorily, omni-directional microphones must be

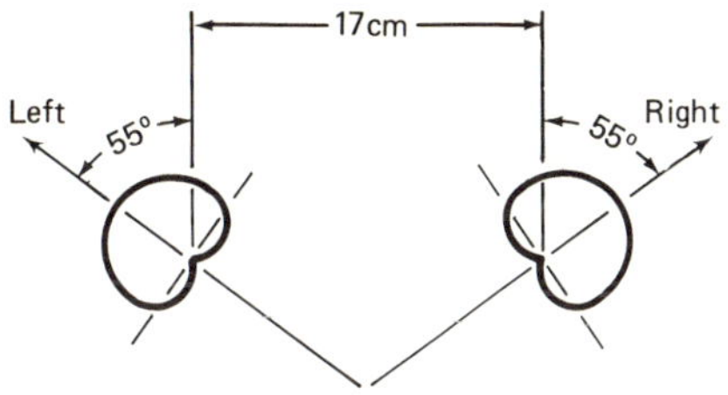

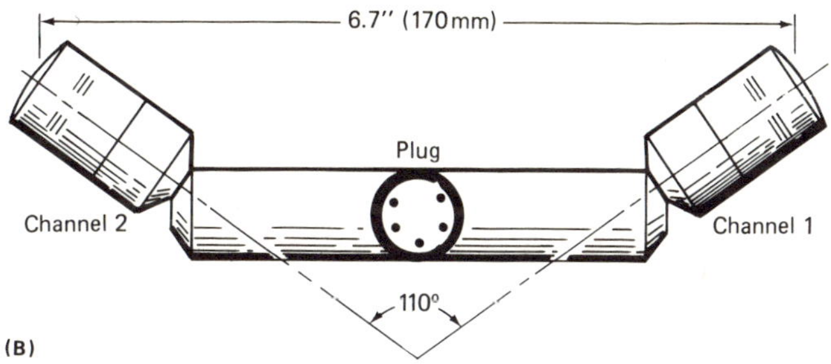

Figure 10-9. The ORTF X-Y array **(A)** and an adaptor for ORTF recording **(B)**.

spaced apart by some distance usually suggested by the musical resources. In Figure 10-11 we show two such spacings: a typical setup for recording a chamber group is shown, as well as a typical setup for orchestral recording. In both examples, the microphone pair is shown responding to a specific instrument in the ensemble. In each case, the distance ratio—and hence the amplitude ratio—is the same at the microphones. However, because of the wider spacing in the orchestral setup, the time delay cues will be greater, thus causing the sound source to be localized farther toward the right loudspeaker, as shown in Figure 10-11C.

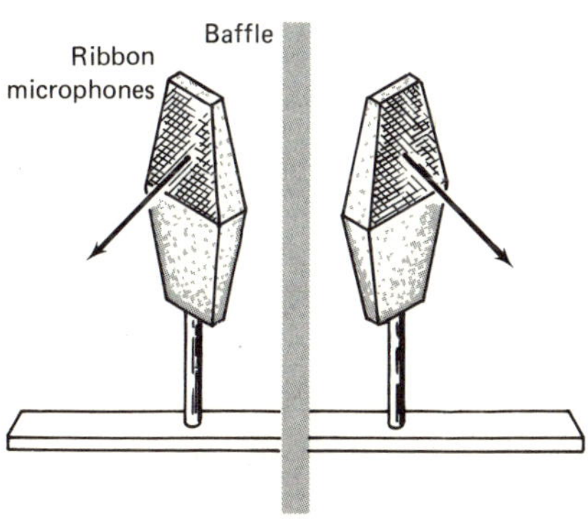

Figure 10-10. A stereo microphone technique proposed by Madsen.

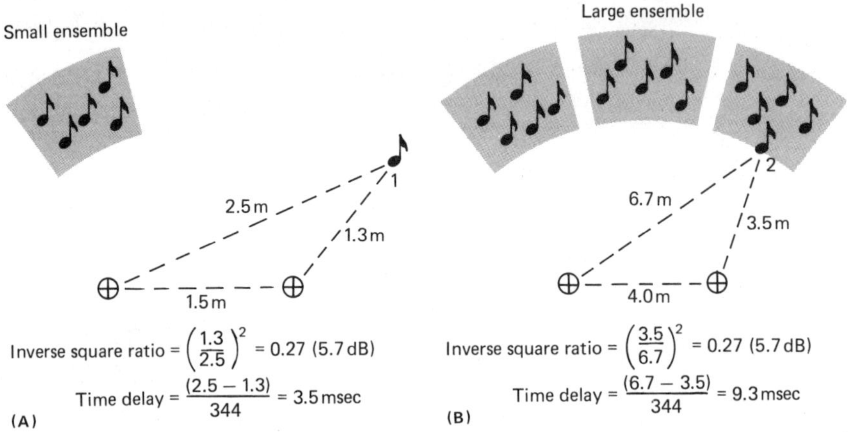

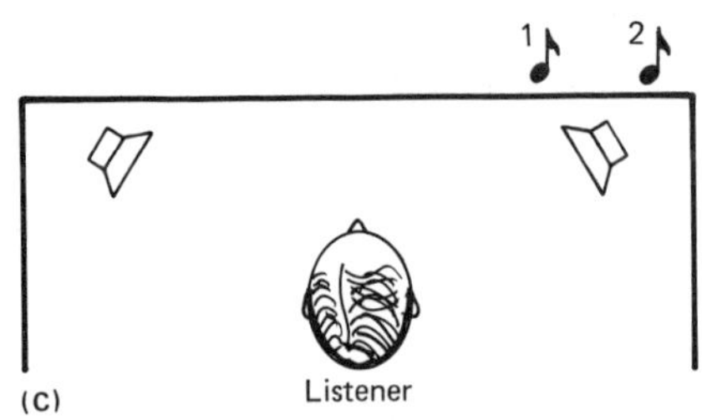

Figure 10-11. Stereo techniques using spaced microphones. Delay and amplitude cues at a pair of microphones for small **(A)** and large **(B)** ensembles; **(C)** the resultant effect on localization.

For wider spacings, there is a tendency for the elements of the ensemble to cluster at the two stereo loudspeakers, as shown in Figure 10-12. When the spacing exceeds, say, three or four meters, many engineers will add a center microphone to the array, panning its output equally into the two stereo signals. This microphone tends to fill in the "hole in the middle" and stabilize localization in the playback array.

DELAY AND AMPLITUDE CUES

In the examples shown in Figure 10-11, we illustrated the roles of amplitude ratio and delay between loudspeakers in establishing localization. In our example, the amplitude ratio remained constant, while the delay factor did not. In figure 10-13A, we give an approximate picture of how localization depends on relative amplitudes at the loudspeakers when a single signal is fed to both. In this case, there is no delay between the two loudspeakers; both are excited at exactly the same time.

In Figure 10-13B, we plot the so-called Haas effect, showing the effect that even a small value of delay may have in directing the sound toward the loudspeaker whose output arrives at the listening position earliest. Delays up to about 30-35 milliseconds will not be heard as such; rather, the sound will be localized at the earlier loudspeaker. However, the delayed loudspeaker may be increased in amplitude by a specific amount to neutralize this image shift caused by the delay. Note that as the delay is increased, up to 10 dB of amplitude imbalance may be required in order for localization to be maintained at the center of the playback array. However, once the delay time becomes long enough to be clearly perceived, the level imbalance will also become apparent to the listener. It is no wonder then that spaced-apart microphones tend to exaggerate left-right relationships, thus often requiring a center-split microphone to stabilize images. The coincident technique operates only on amplitude cues, and accordingly produces an even spread of images across the playback array.

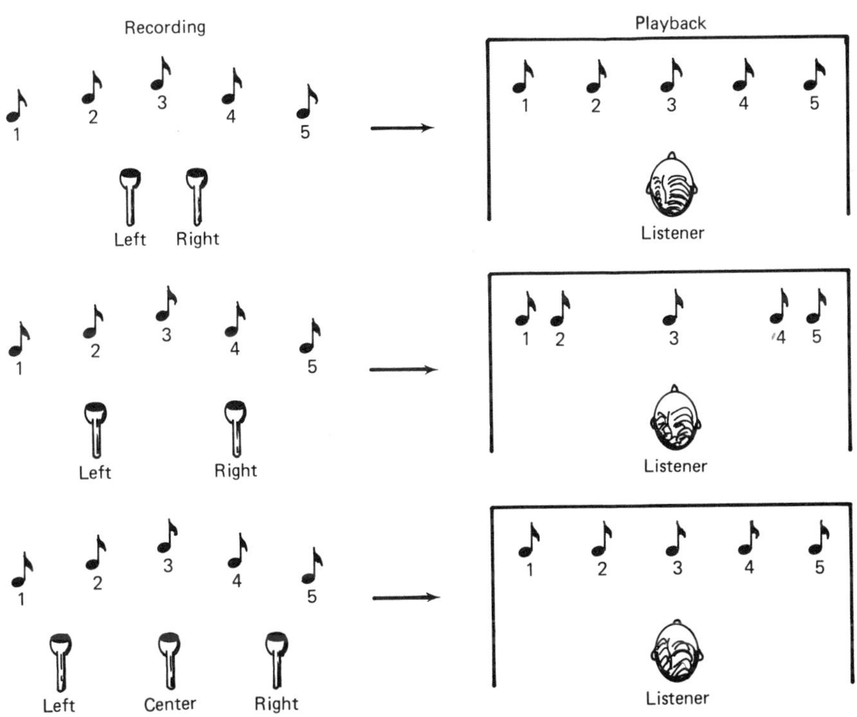

Figure 10-12. Image stability with spaced microphones. As the microphones are moved father apart, there may be a tendency for a "hole in the middle" to appear. A bridged center microphone will usually cure this.

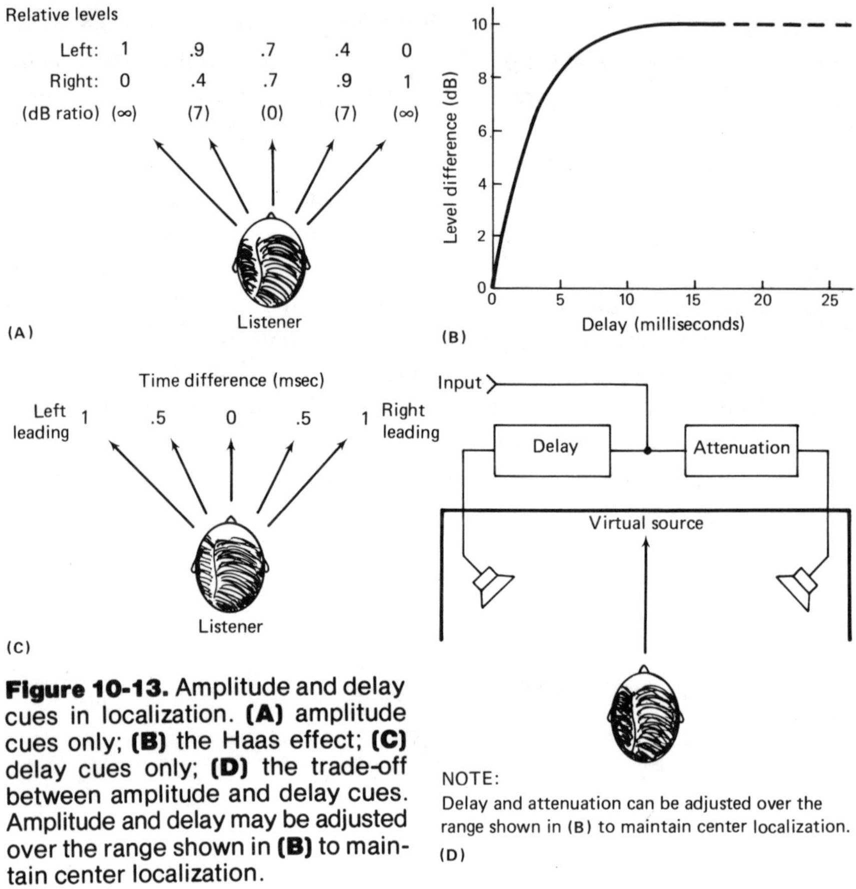

Figure 10-13. Amplitude and delay cues in localization. **(A)** amplitude cues only; **(B)** the Haas effect; **(C)** delay cues only; **(D)** the trade-off between amplitude and delay cues. Amplitude and delay may be adjusted over the range shown in **(B)** to maintain center localization.

USE OF ACCENT MICROPHONES

For the most part, commercial recordings of classical music make use of many microphones. Usually, the overall pickup is provided by a spaced-apart left-center-right group or by a coincident or quasi-coincident pair. So-called accent microphones are then used sparingly to highlight individual sections or instrumental soloists. A typical setup is shown in Figure 10-14. The various accent microphones are panned into the stereo array at positions approximating their actual location in the orchestra.

A particular problem with the use of accent microphones arises from the fact that their response *anticipates* in time the response of the main pickup array. In Figure 10-14, the dotted lines indicate the delay path to the main array. In many typical setups, this delay path is not taken into account, and even a small amount of an early signal from an accent microphone can disturb natural perspectives. Heavy-handed usage of too many accent microphones has a tendency to render the perceived sound stage flat and two-dimensional, since it can swamp out the delicate balance of direct sound and early reflection picked up by the main array.

Onc way to alleviate the problem is to delay the accent microphones by an amount such that the sound reaching the listener from those microphones arrives no sooner than that from the main array, or even a few milliseconds later. When this is done, the Haas effect no longer brings the signals from the accent microphones to the fore, and the balance engineer experiences less difficulty in establishing a pleasing musical balance.

In many ways, the capability of multi-track recorders is antithetical to two-channel stereo recording. Many a recording has been ruined because an engineer took advantage of this capability and proceeded to record each individual section of an orchestra on a different track. An elegant solution to multi-channel orchestral recording has evolved at the hands of the British Decca recording teams, and it is shown in Figure 10-15. The key is in recording a group of stereo pairs, with each pair maintaining its own stereo integrity.

The first stereo pair represents the main pickup array—usually a left-center-right group of three microphones placed on a "tree" some four to five meters above the floor, just behind the conductor's podium.

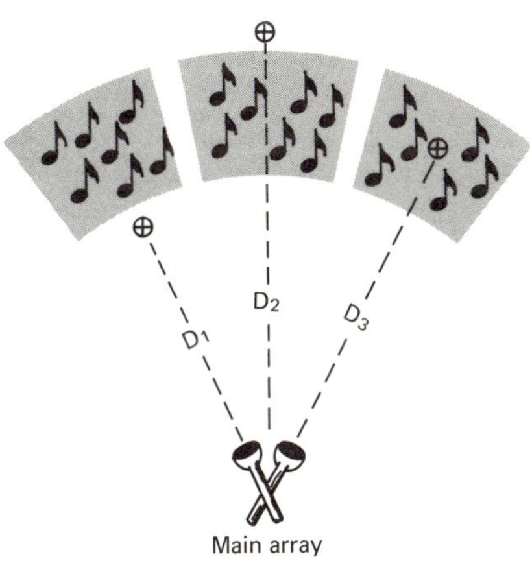

Figure 10-14. The use of accent microphones. For best effect, accent microphones should be delayed relative to the main microphone array, as shown here.

NOTES:

⊕ = Accent microphones at various distances, D_N, from main array

$$T_N = \frac{1000 D_N}{344}$$

T_N = Delay time (msec)

D_N = Distance to accent microphone (meters)

N = 1, 2, 3...

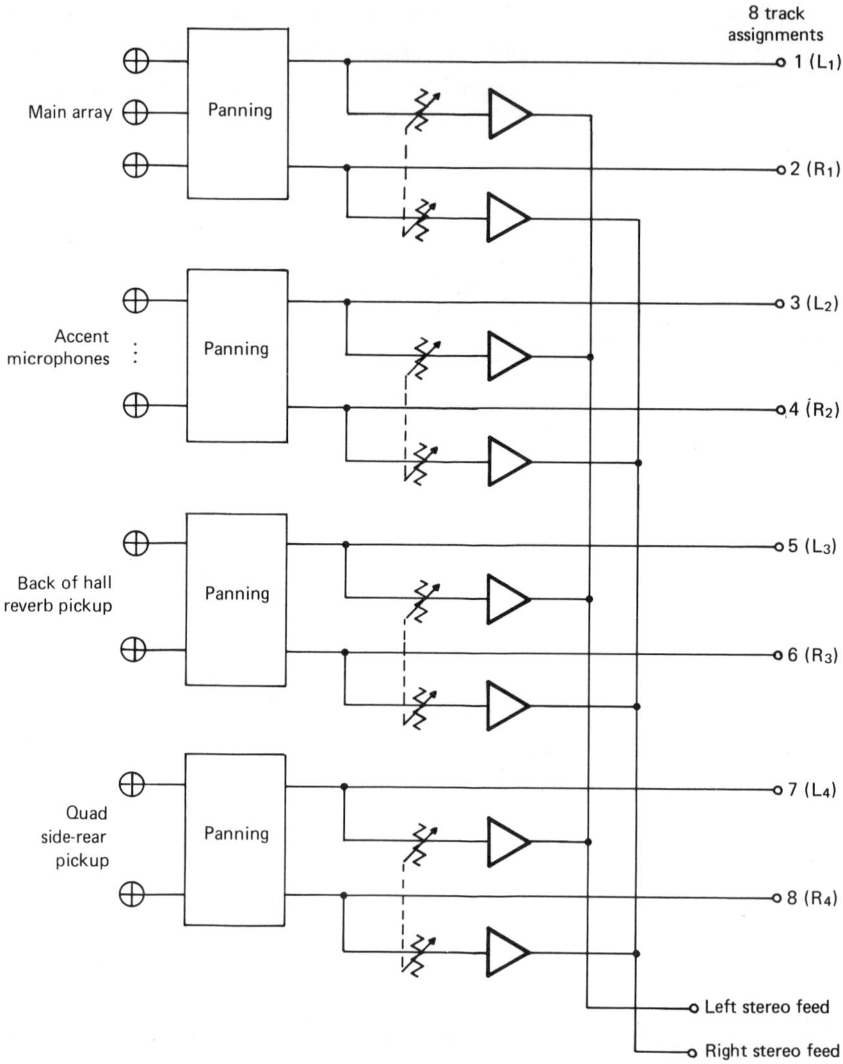

Figure 10-15. The use of multiple stereo pairs in classical recording.

The second pair is the two-channel reduction of the accent microphones, all individually panned to their proper location in the stereo array.

The third pair provides reverberation from the back of the hall. Its role in the final mix is a subtle one, highly dependent on the acoustical nature of the hall.

The final pair is dedicated to the left-back and right-back channels of a quadraphonic array. These would be fed from a pair of microphones located toward the left and right sides of the hall and not more than 30-35 milliseconds delayed through their physical position from the main pickup array. This pair of channels would be used only in the event of a quadraphonic release of the program.

While these eight channels are being recorded, a simultaneous 2-channel mix-down is being recorded. If all goes well, the 2-channel recording can be used for the production of records. However, if subtle balances need to be redone, the 8-track tape may be remixed down to a new 2-track version. In such a post-production mix-down, the 8-track tape would be manipulated as 4 stereo pairs. Suppose the producer, or conductor, felt that a particular instrument had not been highlighted enough via its accent microphone during the original session. On remix, the accent pair of channels would be raised in level, and the proper balance restored. (Of course, the entire ensemble of accent microphones would be raised in the process—but in all likelihood the other accented instruments or sections would be pretty much in the background.) Similarly, the remix engineer could "thin-out" the overall texture by reducing the amount of the reverberation channels in the final mix.

In general, the approach favored by Decca represents a happy combination of multi-channel flexibility and traditional values in classical orchestral recording. Above all, it serves the musical demands in a very effective way.

SUBJECTIVE EVALUATION OF SPACED-APART MICROPHONE TECHNIQUES

Just as some engineers are avid devotees of coincident or quasi-coincident techniques in stereo microphone placement, others favor the spaced-apart approach. As a general observation, the coincident approach preserves geometric relationships in the performing group better than a spaced-apart array does; the listener is better able to pinpoint the positions of individual instruments. By comparison, the spaced-apart approach often conveys a heightened sense of the dimension and characteristics of the recording environment itself. Many listeners favor what they call a "warmer," less analytical, sound through the use of spaced-apart microphones.

In ideal environments, there is little doubt that coincident or quasi-coincident techniques offer a high degree of realism. In less-than-ideal environments, the spaced-apart technique often helps, through exaggeration of time delay cues, to create a sense of larger space.

Finally, we note that the spaced-apart approach results in a somewhat wider listening area than the coincident techniques do, due to presence of time delay cues in the recording process which counteract the amplitude offset which occurs when a listener moves off-axis.

BINAURAL RECORDING

While the artificial head is rarely used in commercial recording, it can, by means of a binaural-to-stereo transformation, create some stunning effects. The technique is shown in Figure 10-16.

If a binaural recording is played directly over loudspeakers, the stereo presentation will be a weak one, due to the presence of excessive leakage around the articifial head. However, these leakage paths can be simulated for a given size and shape of artificial head

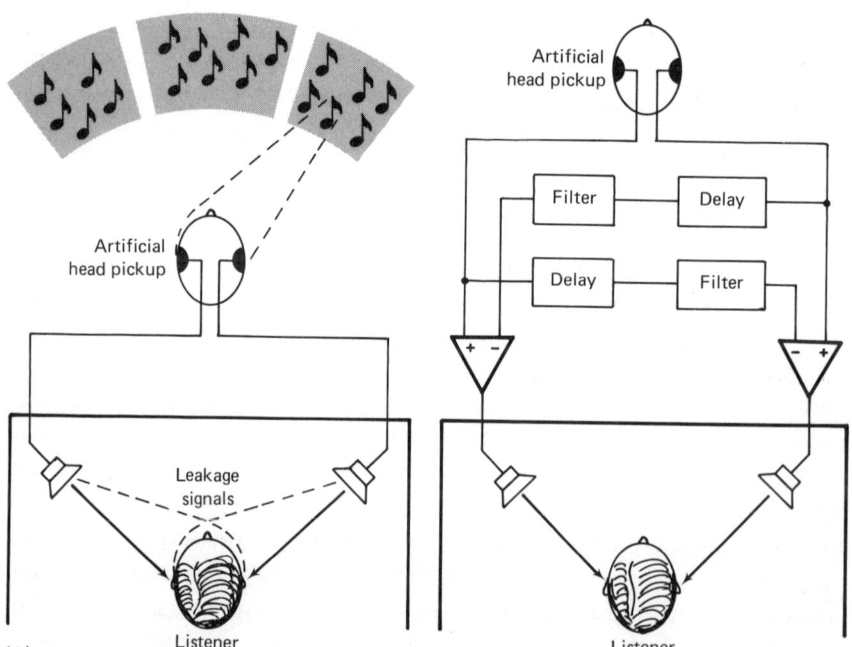

Figure 10-16. Binaural-to-stereo transformation. **(A)** a binaural transmission reproduced in stereo; **(B)** a binaural transmission "corrected" for stereo transmission.

and can then be cross-fed in opposite polarity to cause effective cancellation of them. When this is done, the stereo effect is quite accurate, conveying not only the directions of sounds within the loudspeaker array but also reverberant information outside the loudspeaker array.

The binaural-to-stereo transformation lends itself quite well to the demands of environmental recording and recording in nature.

THE PRESSURE RECORDING PROCESS

In recent years, interest has grown in what has become known as the pressure recording process. The name derives from the fact that the microphone is placed very close to a wall or plane boundary, where it senses only sound pressure (the air particle velocity being effectively zero due to the rigid boundary).

The microphone is located, with respect to the plane, in such a way that its normal zero-degree incidence direction faces the plane and is very close to it, perhaps within a few millimeters. This placement of the microphone reduces the effects of multiple paths to the microphone, as discussed in Chapter 8, and it also significantly diminishes the normal on-axis rise in high-frequency response relative to the random incidence response, as discussed in Chapter 7. The method accomplishes this by preventing sounds from entering the microphone via the usual off-axis paths, as shown in Figure 10-17.

While both these effects are beneficial, the proponents of the pressure recording method point to the equalizing of zero and random incidence response as the major advantage of the method. What this means in practical terms is that the random incidence response may be equalized as the music demands and yet not have the zero-incidence response overly bright. This property of the method is also shown in Figure 10-17.

The proponents of the process recommend fairly-wide spacings for ideal results. These spacings (up to six meters) may be necessary if the microphones are operated at the natural wall boundaries of a large studio with players normally deployed throughout the room. However, with the use of specially constructed flat baffles with microphones located at their centers, such large spacings and large distances from the ensemble are not necessary. As we pointed out in our initial discussion of spaced-apart techniques, the wider spacings of a stereo pair are apt to give rise to unstable center images. The

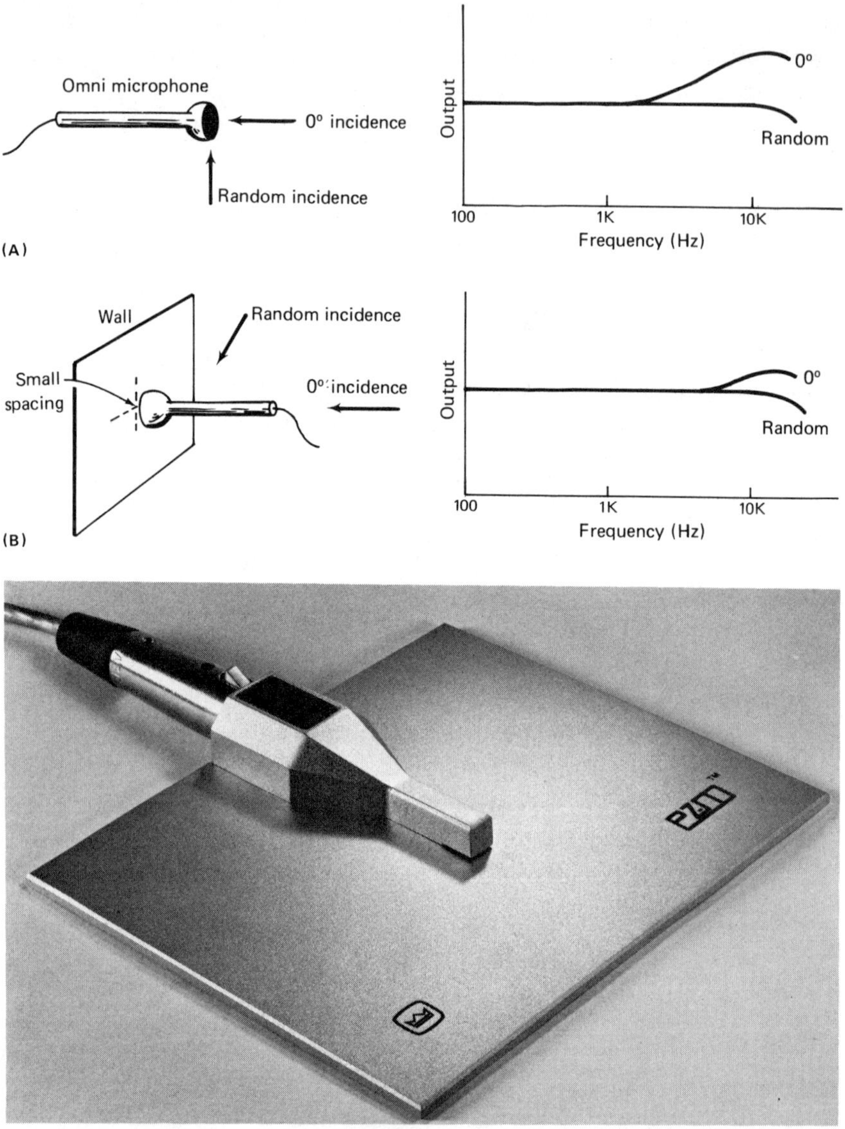

Figure 10-17. The pressure recording process. **(A)** the high frequency difference for normal and random incidence with a typical omni-directional microphone; **(B)** minimized high frequency difference with an omni-directional microphone located close to a wall **(C)** a commercially available PZM (pressure zone microphone) *(Crown International photo).*

pressure recording process is no less immune to this effect than any other spaced-apart technique.

CONSIDERATIONS OF MONO COMPATIBILITY

It should be clear by now that many of the time and amplitude ratios between a stereo pair which make for pleasing stereo will not necessarily result, upon addition, in a balanced mono signal. For many applications, this constraint may not be important; the producer may opt for an ideal stereo mix while letting the mono summation fall where it may. Of course, out-and-out cancellations due to out-of-phase components between the channels are to be avoided, but comb-filter effects due to normal variations in microphone distances from a given source can usually be tolerated. More likely, there will be an ensemble of such comb-filter effects due to the multiplicity of spacings of microphones relative to the many instruments, and the various peaks and dips will not coincide. Thus, the overall mono frequency spectrum will be a smooth one.

A good way to detect potential compatibility problems is to view the stereo pattern on an oscilloscope, as shown in Figure 10-18. If the left and right signals are connected respectively to the vertical and horizontal inputs, then the 'scope display—known as a *Lissajous pattern*—will indicate the relative mono compatibility of the stereo signal.

One reason for the popularity of coincident stereo microphone techniques in Europe is the concern which so many state-operated broadcast systems have for mono playback.

Just how important is mono compatibility in an age which places such strong emphasis on stereo as the normal mode of playback? The answer is that ideally, the integrity of the stereo program should never be sacrificed for the sake of mono. If a balance engineer alternately monitors the program in stereo and mono, then it is likely that serious problems in mono compatibility will be avoided. At the same time, his attention to the stereo balance must result in its optimization.

QUADRAPHONIC MICROPHONE TECHNIQUES

While multi-channel reproduction is not new, it was not until the late sixties that serious attention was given to it as a potential commercial

Figure 10-18. Mono compatibility in the M-S system. When the L and R signals are combined for monophonic use **(A)**, the S components cancel, leaving only the M component. **(B)** Lissajous figures as an indication of compatibility.

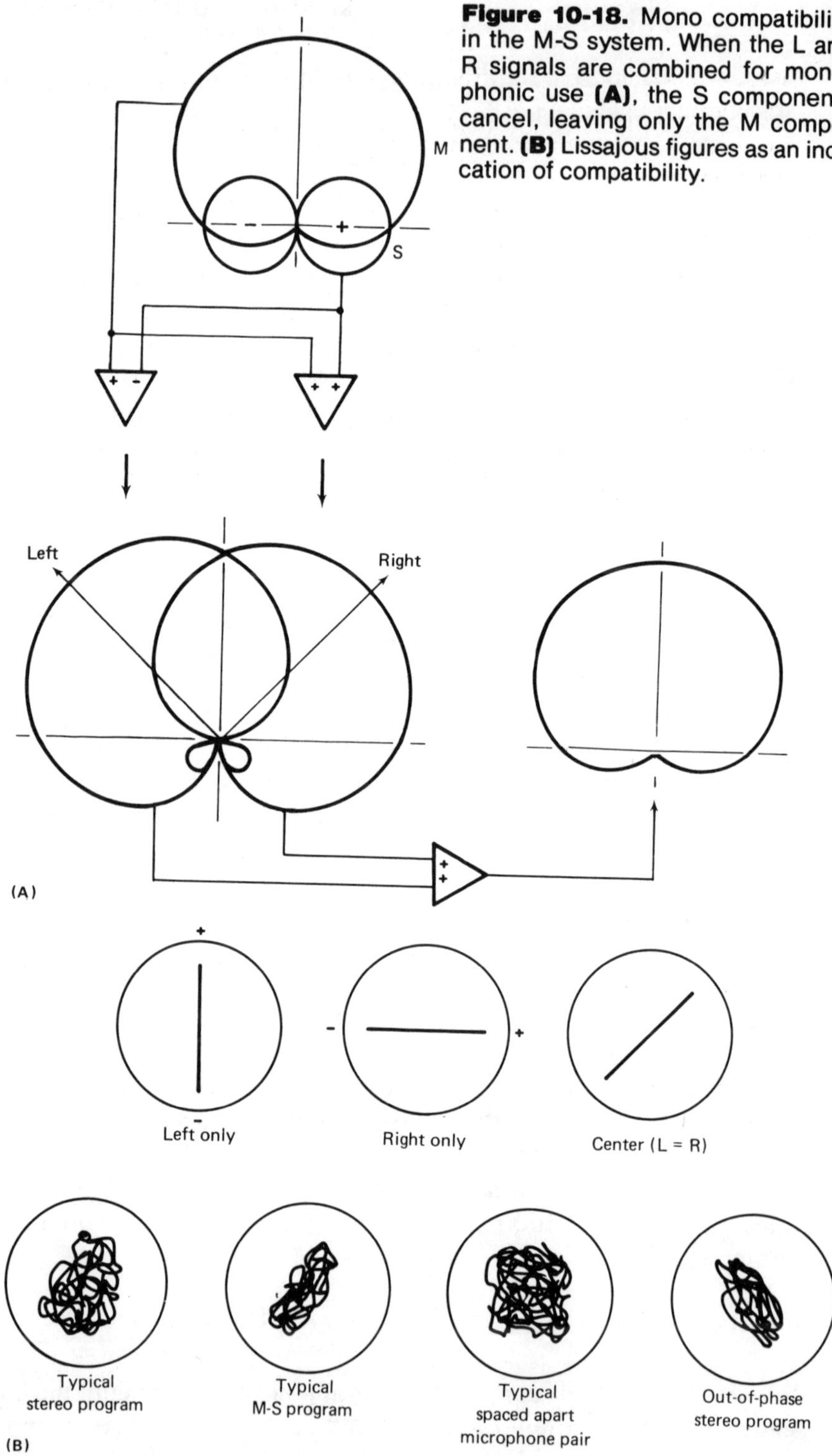

medium. While quadraphonic sound has, so far, failed to attain commercial acceptance (some would say miserably so), a great deal has been learned in the areas of psychoacoustics and sound field synthesis.

As with stereo, it is possible to make quad recordings from an assemblage of tracks, each assigned to a single instrument or to a single section, simply by panning these tracks in wrap-around fashion about the listener. The quad panpot aids the engineer in making these positional assignments. Details of this approach are shown in Figure 10-19.

The approach lends itself to the same excesses of signal processing (artificial reverberation, time delay, etc.) which plague much of

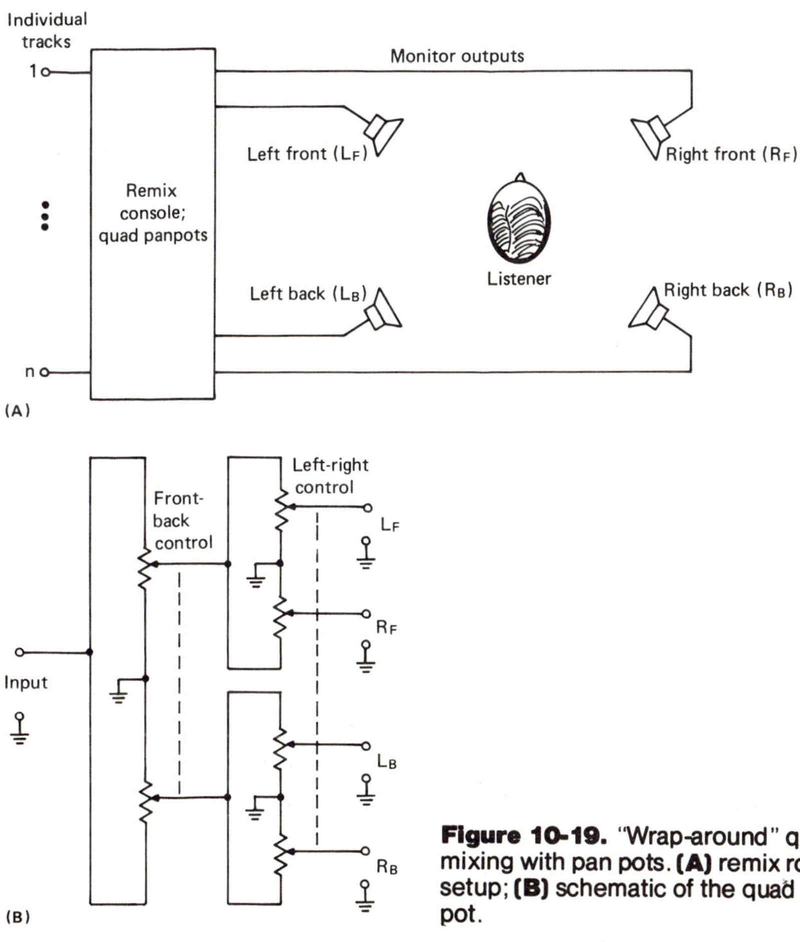

Figure 10-19. "Wrap-around" quad mixing with pan pots. **(A)** remix room setup; **(B)** schematic of the quad pan pot.

popular stereo recording. A far more ambitious goal of quad is to produce a sense of the space perspective of the original recording site. This can be done with spaced-apart microphones, as well as with coincident microphones. Figure 10-20 shows a typical setup in which a forward-oriented pair of microphones picks up the orchestra in the normal stereo fashion. A back pair of microphones picks up essentially reverberant information and is directed to a side or rear pair of loudspeakers.

In setting up such a system, the aim would be to present a normal stereo pickup over the front loudspeakers, while a sense of reverberation and space is presented over all four loudspeakers. Delayed reproduction of the direct sound by the back microphones, provided it does not exceed a 35-40 millisecond amount, will be useful in creating a sense of space. But, if the back microphones are located more than about 12 meters (40 feet) from the front microphones, then the discrete time delay introduced by the back pair picking up the ensemble directly would be confusing and disturbing to the listener.

The choice of microphones for the back pair would depend upon the specific acoustical character of the recording environment. In live halls, it might be necessary to use cardioid patterns, suitably aimed in order to preserve a good balance between reverberant information and direct sound. In rooms which are not too live,

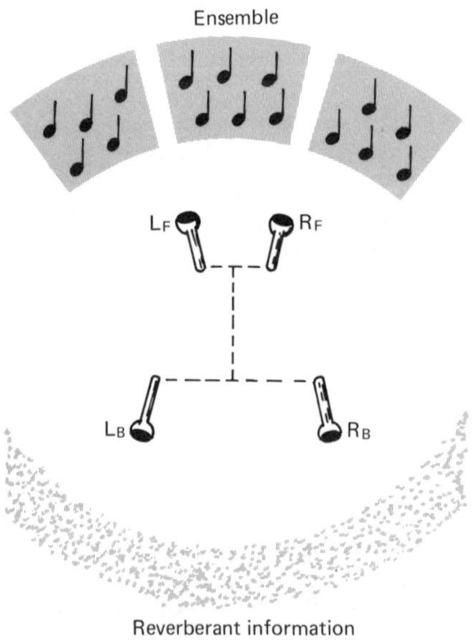

Figure 10-20. An approach to quad microphone technique. The aim here is to capture the ambience of the recording hall for presentation over side or rear loudspeakers.

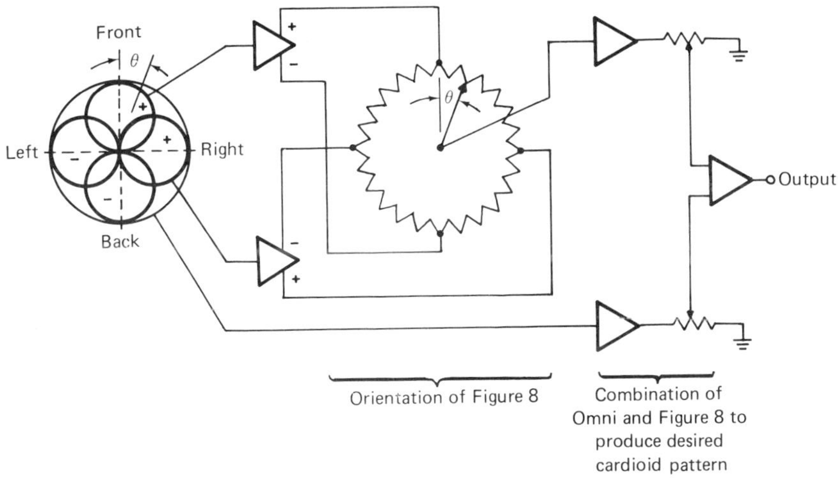

Figure 10-21. A "steerable" microphone. The two figure-8 elements are combined to produce another figure-8 pattern at any desired direction in the horizontal plane. The resulting pattern is then combined with the omni-directional pattern to produce the desired cardioid pattern.

cardioids may be faced toward the back of the room, or omnis may be used.

Coincident techniques are very useful in quad recording, and in order to develop this discussion we will analyze a simple *steerable microphone array*. Figure 10-21 shows an extension of a normal coincident pair, consisting of one omni-directional element and a pair of bi-directional elements, oriented at right angles. If these three microphones are clustered closely together, then their outputs may be combined to produce a cardioid response oriented at any angle in the horizontal plane. A three-channel recording may be made with the array, and it may be played back in such a way as to duplicate the response of a cardioid (or hyper- or super-cardioid) oriented in any direction in the horizontal plane. The two bi-directional patterns can be combined to produce a figure-8 pattern in any direction, θ, and the resulting figure-8 pattern can be mixed with the omni component to produce any desired cardioid pattern.

There are quadraphonic microphones available from some manufacturers. These are assemblies of four cardioid elements oriented in the horizontal plane with some flexibility in the angular positioning of their elements.

A further extension of the coincident concept leads to the four-microphone array shown in Figure 10-22. Here, we have added a third bi-directional element oriented in the vertical plane. The three figure-8 elements may be manipulated to produce a resultant figure-8 pattern oriented at any horizontal angle, θ, and vertical angle, φ, in space. The resulting pattern may then be combined with the omni pattern as before, and a cardioid may be resolved at any direction in space.

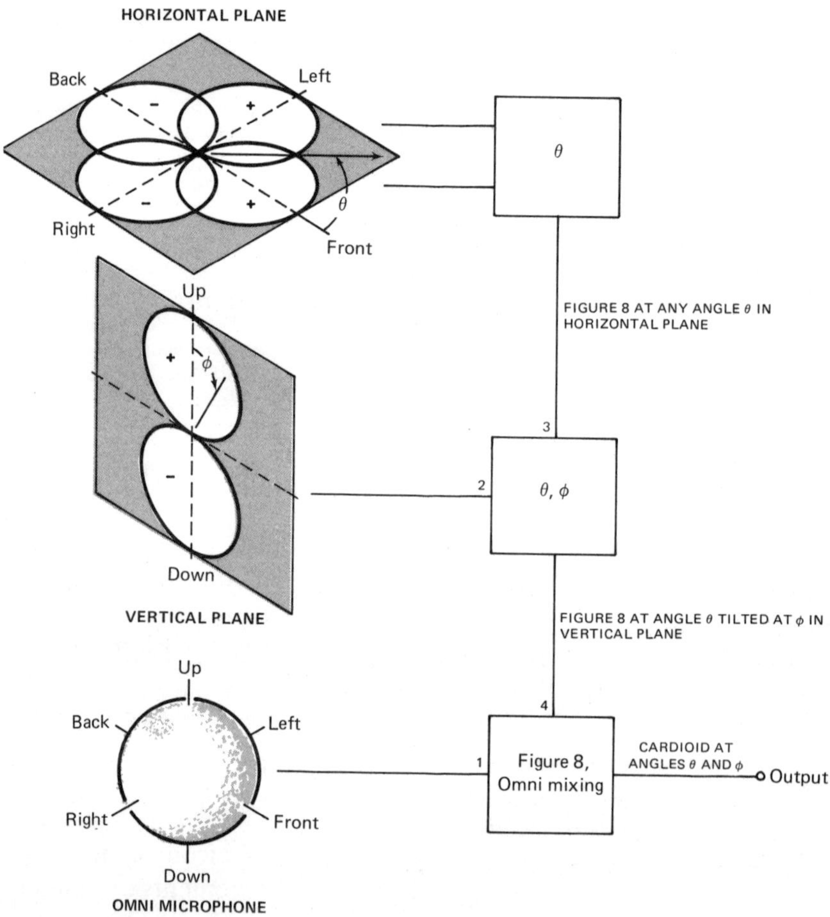

Figure 10-22. A three-dimensional steerable microphone array. An additional vertically-oriented figure-8 microphone is added to the array of Figure 10-21 to create any desired three-dimensional orientation.

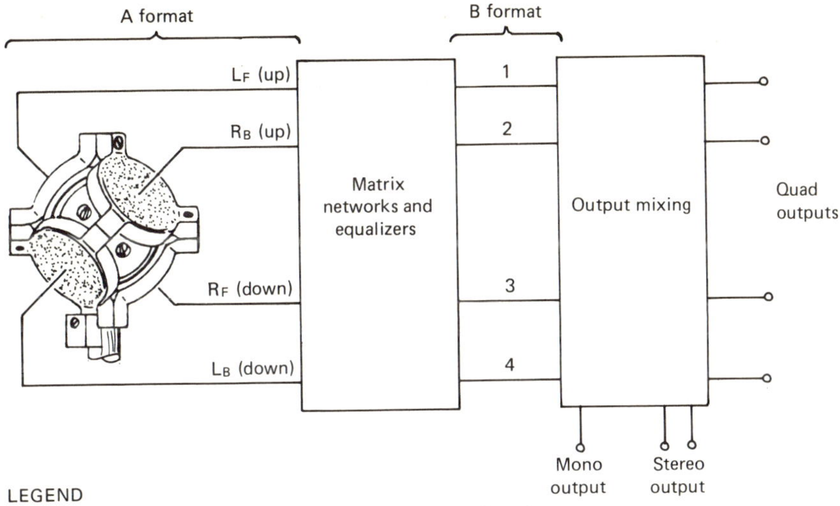

LEGEND

1 = Omnidirectional (pressure) component = $L_F + R_B + R_F + L_B$
2 = Bidirectional (up-down) component = $L_F + R_B - (R_F + L_B)$
3 = Bidirectional (left-right)component = $L_F + L_B - (R_F + R_B)$
4 = Bidirectional (fore-aft) component = $L_F + R_F - (L_B + R_B)$

Figure 10-23. The Sound Field Microphone is a practical embodiment of the technique illustrated in Figure 10-22.

THE SOUND FIELD MICROPHONE

It is quite difficult to get the four microphones of Figure 10-22 close enough together to be truly coincident over the audio frequency band. A better approach has been worked out by Calrec Audio, Ltd., of England, and it is based upon an assembly of four modified cardioid elements which are oriented parallel to the faces of a regular tetrahedron. The positioning of the elements is as shown in Figure 10-23. They are of the single diaphragm type, and the complex paths behind each element have been worked out to provide accurate pattern control over the bulk of the frequency range. Since the array is symmetrical about any one element, the response will be equal for all four. A supporting stem and supports for the wind screen will only affect this symmetry slightly.

The direct output of the four capacitor elements is referred to as the "A-format." After appropriate mixing and equalizing, these signals are converted to the form shown in Figure 10-22 and called the "B-format." It is from the B-format that the signals can be manipulated to produce the following:

1. A steerable mono output, as described in Figure 10-21.
2. A stereo pair whose effective angle, vertical tilt and rotation can be manipulated. The ratio of *forward* or *vertical* component in the mix can be altered to offer more control over room reverberation characteristics.
3. A quadraphonic set of outputs whose effective direction and tilt can be manipulated. The vertical component can be independently manipulated for added control of room reverberation characteristics.

Through its ability to cover all three dimensions of space, the sound field microphone optimizes the use of four recording channels for a single-point pickup. Its utility extends far beyond that of a quadraphonic microphone; its outputs interface with more complex playback arrays, known generally by the term *ambisonic,* which provide for the presentation of vertical, or "sound-with-height," information.

THE Q-BIPHONIC TECHNIQUE

The Victor Company of Japan has developed a dual artificial head technique called Q-biphonic. Details are shown in Figure 10-24. Both heads face forward, separated by about 0.5 meters. There is an absorptive baffle between them, and this causes the heads to be, respectively, forward- and back-oriented. The outputs of both heads are passed through a binaural-to-stereo transformation and played over an array of four loudspeakers. The accuracy of the method in preserving the directional integrity of a sound source as it moves about the artificial head array is shown in Figure 10-24B. Here, the relative size of the circles indicates the overall accuracy of direction perceived by a listener located in the center of the loudspeaker array.

SUMMARY

In this chapter we have seen how varied the techniques of stereo recording can be. There is no single way which is best or which can always guarantee good results. Superb orchestral recordings have been made with as few as two microphones and as many as 24!

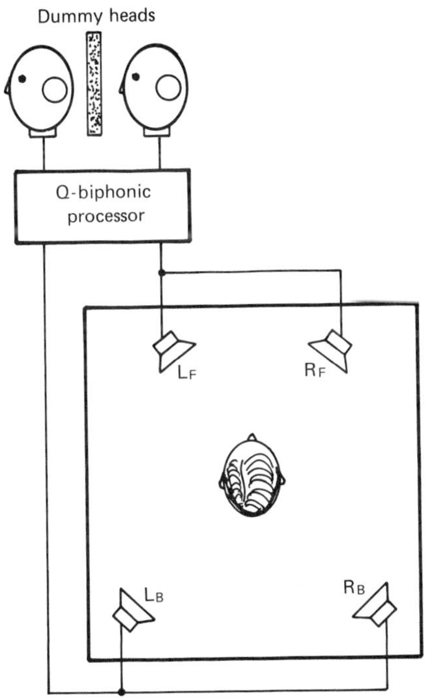

(A)

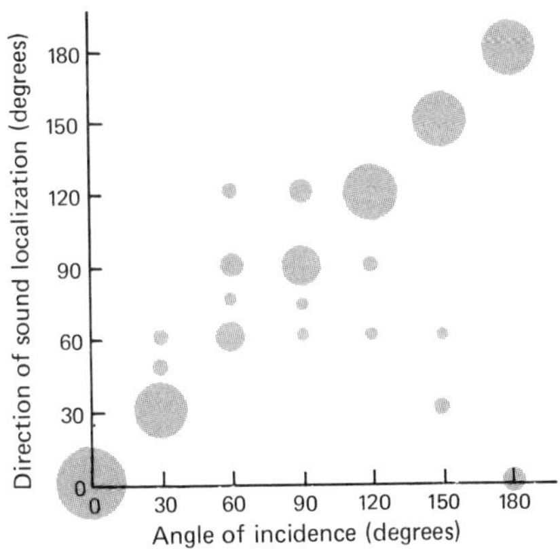

(B)

Figure 10-24. The Q-biphonic system. **(A)** system block diagram; **(B)** effectiveness of the system. The size of the circles indicates the relative scores in localization tests.

Our advice to the neophyte is to begin with simple resources—and move on only after they have been mastered. In particular, the various coincident techniques should be explored first. They have the virtues of demanding relatively little in the way of equipment, and they teach the fundamental "gospel according to Blumlein." Many a seasoned engineer has returned to the simplicity of coincident techniques after years of experience with multi-microphone spaced-apart techniques.

CHAPTER XI

Speech and Music Reinforcement

INTRODUCTION

A frequent problem with sound reinforcement systems is the acoustic feedback loop which may exist between the input and output of the system. In fact, much of the design of these systems is aimed at minimizing the interaction between input and output. Our study will include both music and speech reinforcement systems operating both indoors and outdoors.

SPEECH REINFORCEMENT REQUIREMENTS

The prime purpose of a speech reinforcement system is to provide intelligibility and realism of amplified speech for those members of the audience who are to be covered by the system. Speeches have been made since the beginning of mankind, and one may well wonder how speakers fared during the centuries preceding electro-acoustical reinforcement systems. Figure 11-1 outlines the evolution of natural speech reinforcement in the environment in a way which will help us understand the role a modern reinforcement system plays. We observe a speaker trying to be heard by a crowd of people on flat land. The taller—and louder—the speaker, the better he might be understood, due to the rapid attenuation of sound as it moves over the heads of the audience at such a low grazing angle. Still, the normal attenuation with distance limited the coverage of even the loudest speakers.

Then, we see the next logical step; the speaker has placed himself on a hillside so that there is an effective line of sight between him and each listener. But we still must speak loudly, and there is little room for subtleties of inflection.

The next step is shown in Figure 11-1C. In the typical classical amphitheater, the speaker adds to his line-of-sight advantage two

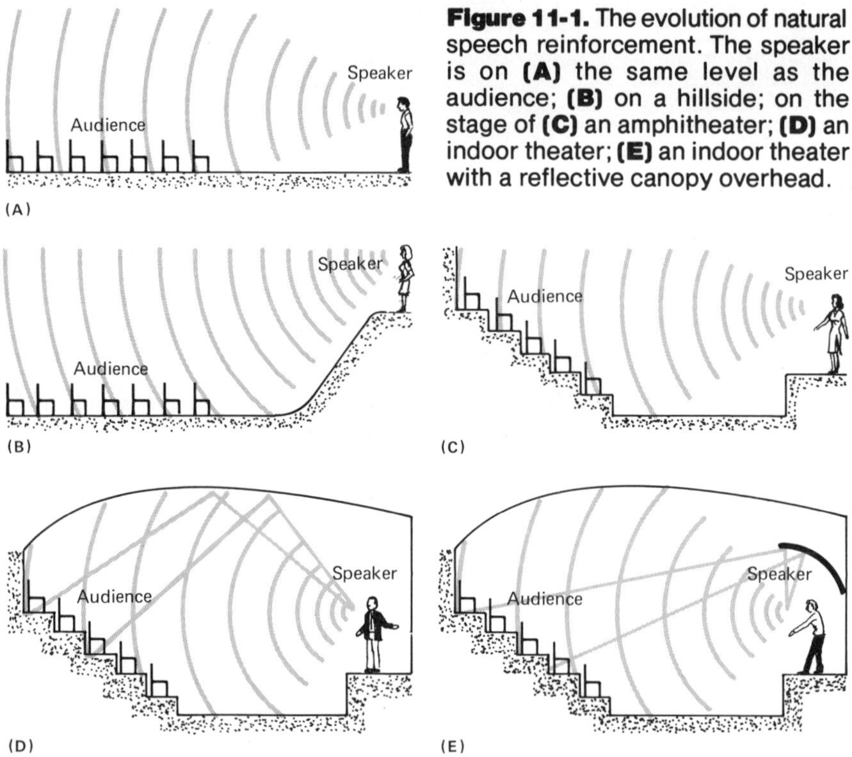

Figure 11-1. The evolution of natural speech reinforcement. The speaker is on **(A)** the same level as the audience; **(B)** on a hillside; on the stage of **(C)** an amphitheater; **(D)** an indoor theater; **(E)** an indoor theater with a reflective canopy overhead.

others: Because the amphitheater is isolated from its surroundings, the ambient noise level is apt to be lower—a great help in improving intelligibility, especially for older listeners, and because the first few reflections of sound are containted within the structural bounds of the amphitheater, they enhance the loudness of the speaker's voice—again a help in improving intelligibility. In the form of the outdoor amphitheater, natural speech reinforcement has been around for many centuries. With actors whose voices project well, and with listeners who are attentive, a wide range of speech levels can be accommodated for dramatic impact, as long as outdoor noises or inclement weather do not interfere.

The next development is to isolate the environment from the elements. The comfort of indoors probably encourage the listener to be more attentive, while the added level due to reverberation, as well as increased noise immunity, helps the actors or speakers to convey a wider range of emotions.

In Figure 11-1E, the final development is shown. Where the speaker can be restricted to a fairly small area, an overhead canopy or shell can be used to increase the directivity of the speaker for greater coverage of the audience area. In this last form, natural speech reinforcement is still with us, with relatively few drawbacks. Properly designed houses of worship, auditoriums and lecture rooms of moderate size may have little or no requirement for electro-acoustical sound reinforcement at all.

Thus, the attributes of good speech intelligibility which have evolved over the historical span shown in Figure 11-1 are:

1. Line-of-sight contact between speaker and listener.
2. Isolation and elimination of interfering noises.
3. Use of natural reverberation as an aid to increasing loudness of the speaker.
4. Increasing the directivity of the speaker toward the audience.

Modern sound reinforcement systems work in concert with these same attributes. In addition, they provide the following:

1. Ability to cover larger audiences.
2. Ability to accommodate soft-spoken or other less-than-ideal speakers.
3. Sufficient electro-acoustical gain to exceed normal requirements.
4. Flexibility to meet changing requirements.

ACOUSTICAL GAIN

As our next step, we examine the concept of *acoustical gain.* Acoustical gain is the difference in level perceived with the system turned off, and the level perceived with the system turned on. *Potential acoustical gain* (PAG) is the maximum gain the system can provide, consistent with some margin against feedback.

The first question to be answered is: how much acoustical gain is necessary? In normal environments, speech communication takes place at the level of about 65-to-70 dB-SPL when talkers are within 1-to-1½ meters of each other. In an especially quiet environment, normal speech can be easily understood by most listeners at distances up to three meters. Thus, a speech reinforcement system should be able to deliver a sound level to the farthest listener corresponding to the typical person-to-person communicating level for the particular environment.

We will now examine a simple reinforcement system operating out of doors and observe those factors which determine its acoustical

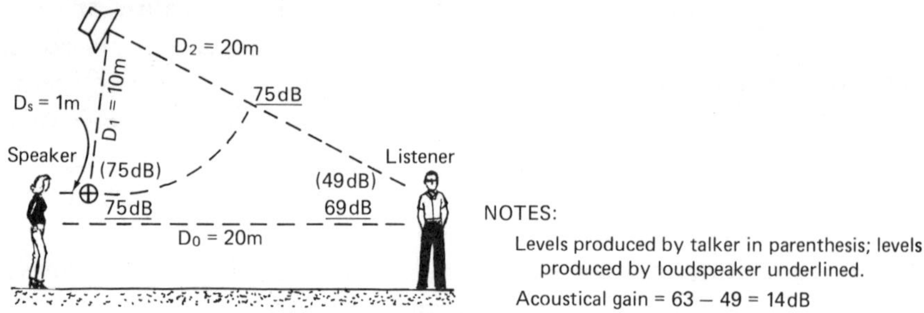

Figure 11-2. An outdoor sound reinforcement system.

gain. In Figure 11-2, let us assume that a performer is located one meter in front of a microhone and is producing average sound pressure levels of 75 dB-SPL at the microphone. A loudspeaker is located 10 meters from the microphone, and the farthest listener is located 20 meters from both the talker and loudspeaker. We further assume that the loudspeaker and microphone are both omni-directional.

With the reinforcement system turned off, the performer will be heard at an average level of 49 dB-SPL; the 26 dB-lower level is due to inverse-square fall-off. When the system is turned on, its gain may be increased, theoretically, until it is unity around the microphone-loudspeaker loop; that is, to the point at which the system's output equals its input. Thus, when the loudspeaker also produces a level at the microphone of 75 dB-SPL (the same as produced by the performer), feedback will begin. Now, when the loudspeaker produces a level of 75 dB-SPL at 10 meters, it will produce, by inverse-square fall-off, a level 6 dB lower at a distance of 20 meters, or 69 dB-SPL. Therefore, the system's acoustical gain is 69 − 49 = 20 dB.

Sound reinforcement systems are never operated this close to feedback. Usually, a safety margin of 6 dB is included in the gain calculations. Thus, in our example, a useful and reliable measure of the potential acoustical gain would be 20 − 6, or 14 dB.

We may generalize the foregoing gain calculations and arrive at a simple method for determining acoustical gain by means of the nomograph shown in Figure 11-3. In this nomograph, inverse-square losses are referred directly to one meter (or one foot), and the decibel values (20 log D) may be read directly from the nomograph as indicated in the following equation:

$$\text{PAG (dB)} = 20 \log D_1' + 20 \log D_0' - 20 \log D_S' - 20 \log D_2' - 6$$

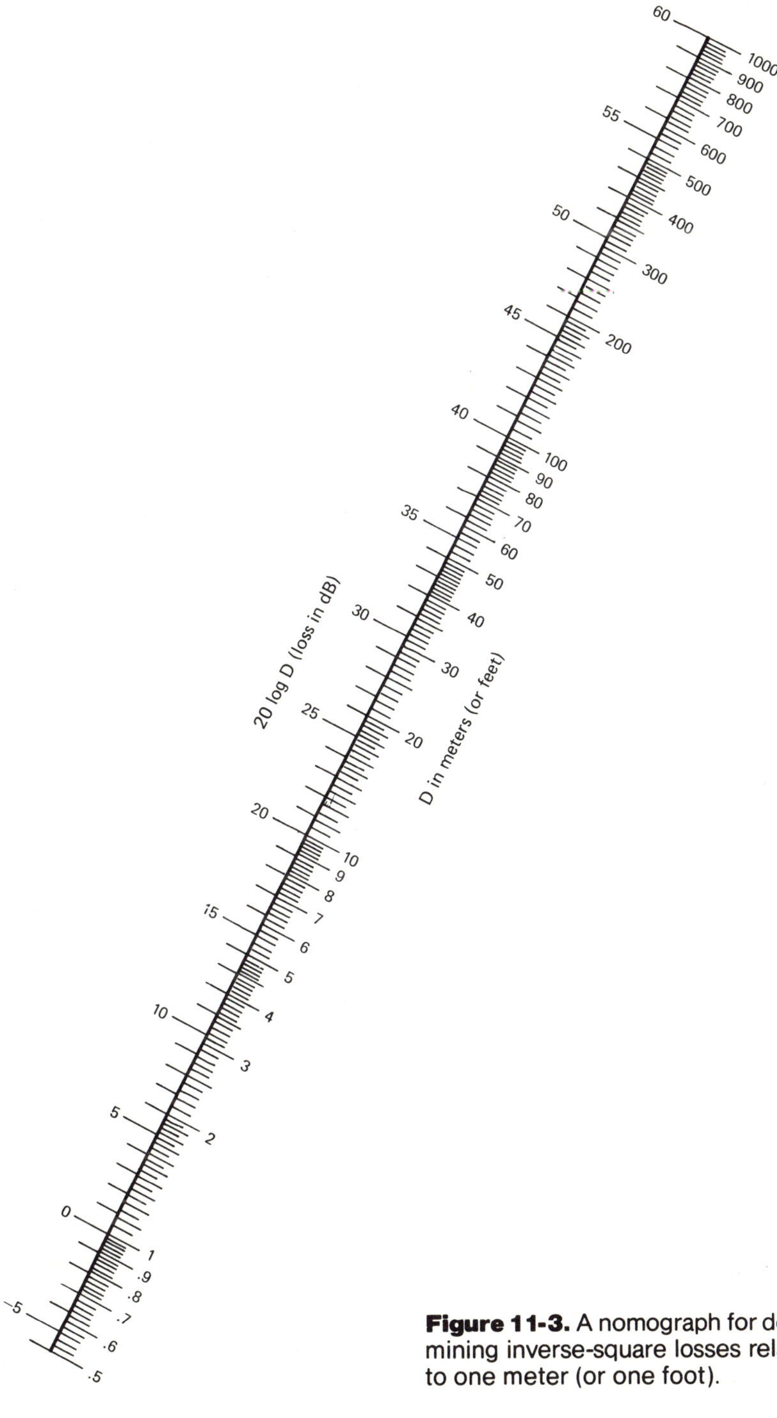

Figure 11-3. A nomograph for determining inverse-square losses relative to one meter (or one foot).

Working our previous example using the nomograph, we have:

$$PAG = 20 + 26 - 0 - 26 - 6 = 14 \text{ dB}$$

The next question we must ask is: *how much* gain is needed? The *needed acoustic gain* (NAG) is that system gain which, at some remote point in the space covered by the sound system, will provide a sound pressure level equivalent to normal speech at a normal communicating distance. The purpose of the sound reinforcement system is, in effect, to bring the performer forward; that is, close to the listener, as shown in Figure 11-4. For most reinforcement systems located in quiet environments, this desired *equivalent acoustic distance* (EAD) will be on the order of about three meters. For systems in noisier environments, the requirement may be as close as one meter.

Returning now to our earlier example, we note that the system could provide 14 dB of acoustical gain. If all of this gain is utilized, the effect is that of transporting the performer forward to a distance of 4 meters from the listener. (Using the nomograph of Figure 11-3, locate the distance D_0 of 20 meters and note the corresponding level of 26 dB. Now, subtract from that level 14 dB, the system's PAG. This corresponds to a level of 12 dB, and we read its corresponding distance, 4 meters.)

We can now define the needed acoustic gain of the system as follows:

$$NAG = 20 \log D_0 - 20 \log EAD$$

It is only necessary that the reinforcement system "bring" the performer closer to the farthest listener by that amount of inverse-square loss which occurs between the equivalent acoustic distance and D_O:

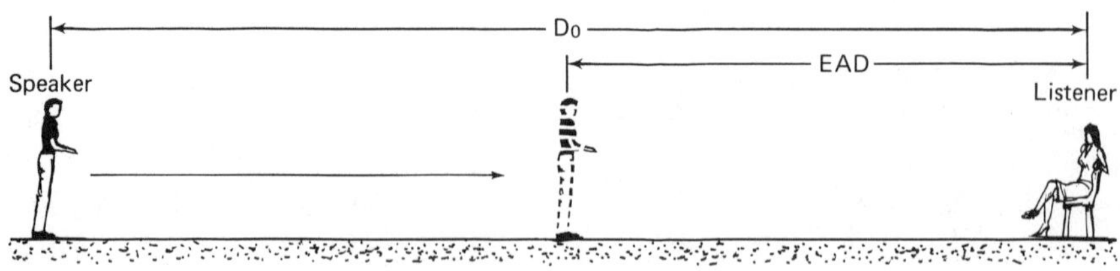

Figure 11-4. Equivalent acoustic distance. One role of sound reinforcement is to effectively bring the performer, through amplification, to within some desired distance from the listener.

In the system under discussion, let us assume a quiet environment and an EAD of 3 meters. Thus:

$$\begin{aligned} \text{NAG} &= 20 \log 20 - 20 \log 3 \\ &= 26 - 9.5 = 16.5 \text{ dB} \end{aligned} \tag{2}$$

Obviously, the system cannot satisfy the requirements; we need 16.5 dB of gain, and the potential acoustical gain is only 14 dB. There is only one sensible solution to the problem: reduce D_S to 0.5 meter. Note from Figure 11-3 that this would add 6 dB to the system's potential acoustical gain, and this will result in PAG exceeding NAG by 20 − 16.5 = 3.5 dB.

In most outdoor speech reinforcement applications, considerable acoustical gain is needed, and it is customary to use fairly short values of D_S. For podium usage, flush mounting, which would normally be preferred for its freedom from interference problems, may not result in a sufficiently low value of D_S. A personal microphone or one with a gooseneck adaptor would probably be the best solution.

THE EFFECT OF LOUDSPEAKER AND MICROPHONE DIRECTIVITY ON GAIN

Continuing our discussion of outdoor sound systems, the PAG of a system may be increased by beneficial use of both microphone and loudspeaker directivity patterns, as shown in Figure 11-5. Normally, the major axis of a directional microphone is aimed at the talker, while the major axis of a directional loudspeaker is aimed at the farthest listener. We can increase the sound attenuation along the D_1 path (and thus increase PAG) by the off-axis attenuation of both the microphone and loudspeaker. With the assumptions made in Figure 11-5, a 9 dB increase in attenuation is indicated. Thus, the joint DI of the loudspeaker-microphone combination is 9 dB.

In practical situations, however, increases up to 4 or 5 dB may be realized by this technique. The reason for this is simply that, as we increase attenuation along the D_1 direction, the role of reflections (they exist in all normal outdoor conditions) becomes more significant.

In a sense, the microphone and loudspeaker create along the D_1 direction their own combined directivity index. This condition exists under all acoustical conditions, indoors as well as out, but many sound system designers do not take it into account.

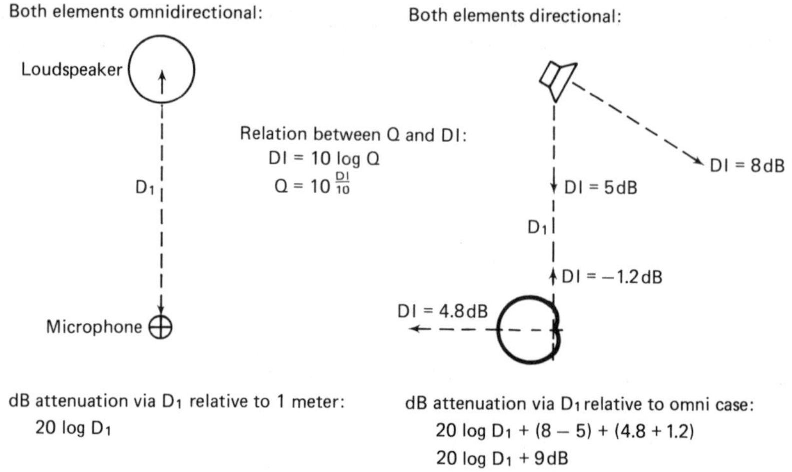

Figure 11-5. Increasing system gain by using directional loudspeakers and microphones. Attenuation along the path, D_1, can be effectively increased by orienting both microphone and loudspeaker so that the line connecting them lies along axes of lesser sensitivity

THE EFFECT OF MULTIPLE MICROPHONES

When more than one microphone is used in a reinforcement system, the potential acoustical gain is reduced. If all the microphones are operating at the same level, then it can be shown mathematically that the potential acoustical gain is reduced *3 dB* with each *doubling* of open microphones. The equation for this is

Gain reduction = 10 log N,
where
N = number of open microphones.

In most such reinforcement applications such as these, there is usually one person speaking at a time. However, for dialogue or discussion, it is desirable to have a pair of open microphones. The reduction in potential gain for a multiplicity of open microphones is shown in Figure 11-6.

When there is a need for many microphones, as for a typical panel discussion, board room meeting, or house of worship, the levels of open microphones may be satisfactorily controlled by an *automatic*

Number of open microphones	Reduction of system gain (dB)
1	0
2	3.0
3	4.8
4	6.0
5	7.0
6	7.8
7	8.4
8	9.0
9	9.5
10	10.0

Figure 11-6. The effect of multiple open microphones on system gain.

microphone mixer, a device which can do the following:

1. Provide speech-actuated gating of individual microphones.
2. Provide overall system attenuation according to the 10 log N rule.
3. Provide for discrimination against false triggering of gating thresholds due to varying ambient noise levels.

A functional diagram of an automatic microphone mixer is shown in Figure 11-7.

ANALYSIS OF AN INDOOR SPEECH REINFORCEMENT SYSTEM

As we discussed in Chapter 8, the influence of the reverberant sound field usually means that the requirements for the acoustical gain of a sound system are relaxed. Moderate reverberation actually assists the

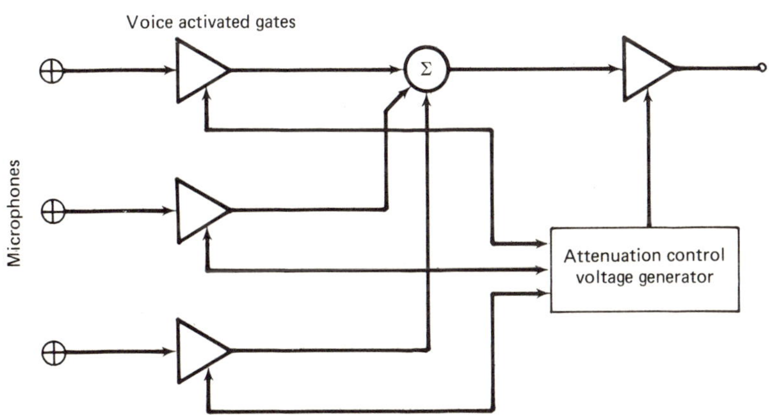

Figure 11-7. An automatic microphone mixer.

talker, and, as we have noted, many small rooms may not need sound systems at all. The equations we developed earlier in this chapter can be modified for the indoor case.

In this section, we will make use of the concepts of directivity factor (Q) and room constant (R). We will not develop these concepts in any rigorous way, since that is well beyond the scope of this book.

Q is a measure of the directivity of a source or receiver. Mathematically, it is a measure of the power put forth (or received) along a given axis, compared to the total amount of power emitted (or received) along all axes. The Q of a loudspeaker will be at a maximum along its major axis, and values of 5 to 8 are common for most loudspeaker arrays used in single-point reinforcement systems. The value of Q along an off-axis direction may be quite low, even less than 1.

ROOM ABSORPTION

The absorption in a room may be calculated from the equation

$$R = \frac{S\bar{a}}{1 - \bar{a}}$$

where

R = room constant
S = boundary area of the room, in square meters, and
$\bar{a}$ = average absorption coefficient of the materials comprising the boundaries of the room.

The average absorption coefficient is calculated from the individual absorption coefficients of the boundary materials as:

$$\bar{a} = \frac{S_1 a_1 + S_2 a_2 + S_3 a_3 + \ldots S_n a_n}{S_1 + S_2 + S_3 + \ldots S_n}$$

This is a tedious calculation, and one that need not be made if the reverberation time of the room is known. In that case,

$$\bar{a} = 1 - \varepsilon^{-0.16V/TS}$$

where

$\varepsilon = 2.718$,
V = room volume, in cubic meters,
T = reverberation time, in seconds, and
S = surface area, in square meters.

The reverberation time is given by

$$T = \frac{0.16V}{-S \cdot \ln(1 - \bar{a})}$$

The reverberation radius, or Critical Distance, D_C, is defined as

$$D_C = 0.14\sqrt{QR}.$$

D_C is that distance away from a sound source where the direct sound field of the source and the reverberant sound field in the room are equal. Critical distance is an important characteristic of a given source in a given room. Because of the pervasiveness of the reverberant field, the attenuation of sound beyond D_C will only be an additional 3 dB (see Figure 11-9).

For our first indoor example, let us consider a *single point* system, where there is a single loudspeaker placed high over the talker's position. Let us make the following assumptions:

1. The microphone is omni-directional, and is located well within the direct sound field produced by the talker.
2. Both the microphone and farthest listener are well within the reverberant field produced by the loudspeaker.
3. The farthest listener is well within the reverberant field produced by the unaided talker.
4. The environment is a quiet one and an EAD of 3 meters is sufficient.

We have diagrammed this system in Figure 11-8, and in Figure 11-9 we show the attenuation of sound with distance characteristic of the talker and loudspeaker array used in this system. Note that the loudspeaker is more directional than the talker; the critical distance for the talker (D_{CT}) is shorter than that of the loudspeaker (D_{CL}).

When the system is turned off, the listener will perceive the sound of the talker attenuated by distance to that level determined by D_{CT}. If the speaker produces a level, L, at the microphone (a distance of

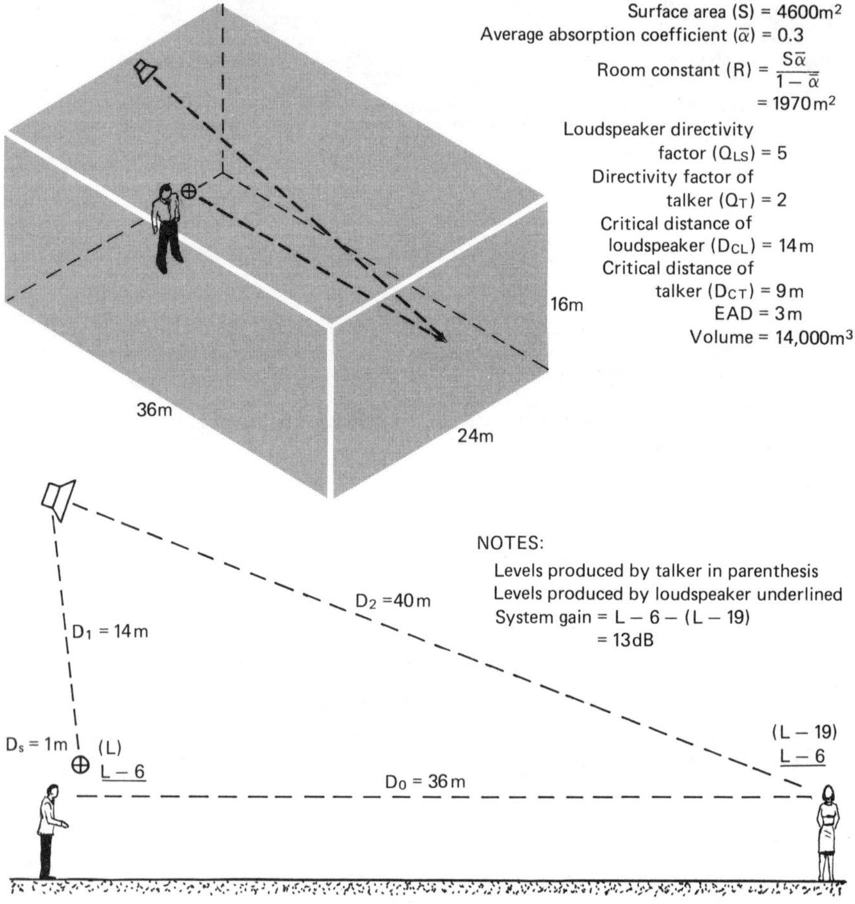

Figure 11-8. An indoor sound reinforcement system.

one meter), then the attenuation at the listener will be:

$$L - 20 \log \frac{D_{CT}}{D_S}, \text{ or } L - 20 \log D_{CT} + 20 \log D_S$$

Now, when the system is turned on, the gain can be increased up to such a point that the loudspeaker produces a level, L − 6, at the microphone (remember the 6 dB safety factor). Since the microphone is in the loudspeaker's reverberant field, it is clear that L − 6 is the level which will exist at D_{CL} and any point beyond in the audience area. Therefore, the difference in level perceived by the listener when the system is turned on is:

$$\begin{aligned} PAG &= L - 6 - L + 20 \log D_{CT} - 20 \log D_S \\ &= 20 \log D_{CT} - 20 \log D_S - 6 . \end{aligned} \quad (3)$$

Note that there is only one useful variable in the above equation. The D_{CT} term is pretty much fixed by the acoustics of the room, and only the D_S term can be effectively manipulated to change the PAG of the system. The reason for there being only one variable is of course our assumption that both the microphone and the listener are in the reverberant field of the loudspeaker.

Solving for PAG using the nomograph of Figure 11-3:

$$\begin{aligned} PAG &= 20 \log 9 - 20 \log 1 - 6 \\ &= 19 - 0 - 6 \\ &= 13 \text{ dB}. \end{aligned}$$

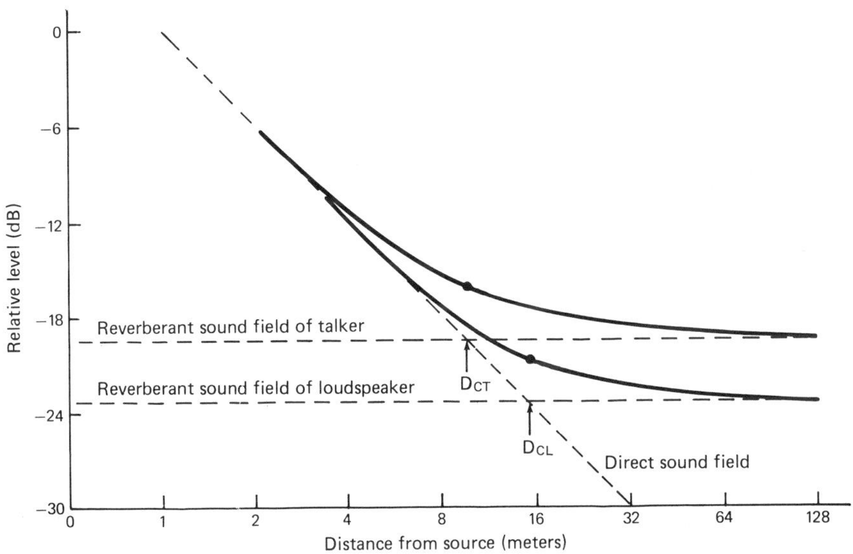

Indoor attenuation dB (as a function of r, distance from source) $= 20 \log \sqrt{\frac{Q}{4\pi r^2} + \frac{4}{R}}$

Q = Directivity factor of source

R (Room constant) $= \frac{S\bar{\alpha}}{1 - \bar{\alpha}}$

$D_{CL} = .14\sqrt{Q_{LS}R}$

$D_{CT} = .14\sqrt{Q_T R}$

Figure 11-9. Attenuation of sound indoors.

As before, we can calculate NAG for our indoor system by noting that we only have to make up the loss in level from EAD to D_{CS}:

$$NAG = 20 \log D_{CS} - 20 \log EAD. \qquad (4)$$

In the present example, D_{CS} is 9 meters and EAD 3 meters; thus, from the nomograph of Figure 11-3:

$$\begin{aligned} NAG &= 20 \log 9 - 20 \log 3 \\ &= 19 - 9.5 \\ &= 9.5 \text{ dB.} \end{aligned}$$

Since PAG exceeds NAG by 3.5 dB, the system will have reserve gain and should be quite stable. Let us note that if we made use of all the potential gain of the system, the effect would be that of moving the talker to within a distance of about 2 meters from the listener.

THE GENERAL INDOOR GAIN EQUATION

The aforementioned example makes the important assumptions that the listener and microphone are in the reverberant field of the loudspeaker, and the gain equation is accordingly simplified. Should this not be the case, we must use Equation (1), modifying the attenuation along each direction by the particular direct-reverberant relationship existing along that direction. Since the attenuation along each path is no longer simply an inverse-square relationship, we will rewrite Equation (1) as follows:

$$PAG = f(D_1) + f(D_0) - f(D_S) - f(D_2) - 6. \qquad (5)$$

We have used $f(D)$ to indicate some *function* of D which will take into account the effects of both direct and reverberant fields on the attenuation of sound in each direction, D, with respect to one meter. Specifically, each direction must be modified as described below:

1. D_1. Attenuation curve based on the critical distance due to the room constant and the joint directivity factor of the loudspeaker-microphone combination acting along their common axis. (See Figure 11-5.)
2. D_0. Attenuation curve based on the critical distance due to the room constant and the directivity factor of the talker.

3. D_S. Attenuation curve based on the critical distance due to the room constant and the joint directivity factor of the talker and microphone along their common axis. (In most cases the D_S term in the equation reduces to a single inverse-square relationship due to the normal short working distances. However, in some theatrical reinforcement systems, footlight microphones may be far enough away from the actors to require a calculation of the critical distance.)

4. D_2. Attenuation curve based on the critical distance due to the room constant and the directivity factor of the loudspeaker along the axis of the listener.

Making meaningful calculations based on these assumptions requires a good bit of knowledge of the acoustical characteristics of the room and the directional characteristics of the loudspeaker array. Furthermore, a thorough designer will perform these calculations for at least three frequencies; typically, 250, 1000 and 4000 hertz. The reason for this is that both the room constant and the various directivity factors involved vary with frequency.

In a similar manner, the NAG equation must be modified:

$$\text{NAG} = f(D_0) - 20 \log \text{EAD}. \tag{6}$$

These calcualtions are not difficult, but they are rather tedious, and engineers who are interested in a more thorough understanding of the principles of sound reinforcement are urged to study those texts which deal with the subject in an analytical way, and to participate in the various "workshop" study programs which are offered in the field by equipment manufacturers and institutes. Our treatment in this book must, of course, be a limited one.

It is appropriate that we add a few notes on the general accuracy of the gain equations presented in this chapter. The equations are based upon simplified statistical models. While we have worked many of these examples out to values of 0.5 dB, we hasten to add that the estimation of gain of an actual sound system will not be so precise. In fact, an agreement between actual and theoretical values within 2 dB is considered excellent.

SOME FACTORS AFFECTING SPEECH INTELLIGIBILITY

Thus far, we have been concerned only with the reinforcement system's capability of delivering a certain level of sound to the listener. In actuality, there are many other factors which determine how effective a sound reinforcement system will be. Some of the more important considerations are:

1. The nature of the talker. Professional speakers do not necessarily talk more loudly than laymen; however, they will enunciate more clearly and pace themselves according to local acoustical conditions. A good speaker will slow down his delivery somewhat in a live room.

2. The nature of the listener. There are both good and poor listeners. Age is an obvious factor, as is interest in what is going on. Higher levels of amplified speech are useful in aiding the interested hard-of-hearing as well as the distracted young person of good hearing.

3. Speech level-to-noise ratio. The exact requirements for signal-to-noise ratio for good intelligibility vary with the level of the speech itself. Generally, a signal-to-noise ratio of about 25 dB represents a limit beyond which a further increase in intelligibility will not result.

4. Direct-to-reverberant ratio. For reverberation times less than about 1.5 seconds, the direct-to-reverberant ratio is of not much consequence in affecting speech intelligibility. As the reverberation time increases beyond 1.5 seconds, the effect of overhang on individual syllables diminishes general intelligibility. Thus, beyond reverberation times of 1.5 seconds, the direct-to-reverberant ratio will materially affect intelligibility.

Figure 11-10 gives practical guidelines for the direct-to-reverberant ratio as perceived by the farthest listener. As the reverberation time increases beyond 1.5 seconds, the direct-to-reverberant ratio must be kept greater.

In the figure, the direct-to-reverberant ratio is expressed in decibels

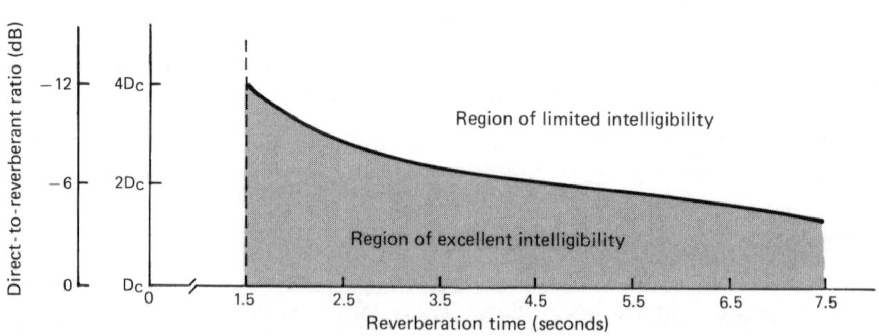

Figure 11-10. Suggested limits for direct-to-reverberant ratio for good intelligibility in sound system design.

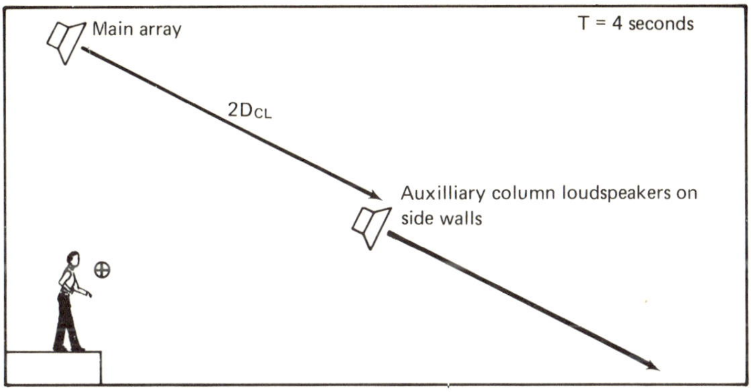

Signal to auxilliary loudspeaker delayed by: $\frac{2D_{CL}}{344}$ seconds. (D_{CL} in meters)

Figure 11-11. The use of an auxiliary loudspeaker to cover the rear of a highly-reverberant room.

as well as in multiples of the critical distance along the axis of the loudspeaker array pointing at the farthest listener.

If a room is excessively long for its reverberation time, then the most remote listeners may be too far into the reverberant field for good intelligibility. An auxiliary set of loudspeakers may be used to cover the rear portion of the room. These loudspeakers, fed at an appropriate power level, effectively increase the direct-to-reverberant ratio for these listeners, enabling them to hear as well as those located toward the front of the room. The use of time delay in the feed to these auxiliary loudspeakers ensures that they will blend into the sound field originating from the main loudspeaker array.

REINFORCEMENT SYSTEM EQUALIZATION

Where there is a concern for good intelligibility, loudspeakers will invariably be equalized for smooth overall response. By no means is equalization only needed for a poorly-arrayed system; even when due attention has been given to all the hardware aspects of a system, there will always be room for extra smoothing of the response through electrical equalization.

In addition to the overall, or *broadband,* smoothing out of system response, *narrow-band* equalization can aid in the increase of acoustical gain through the control of those narrow areas in the response curve where feedback has a tendency to take hold. In fact, the safety factor

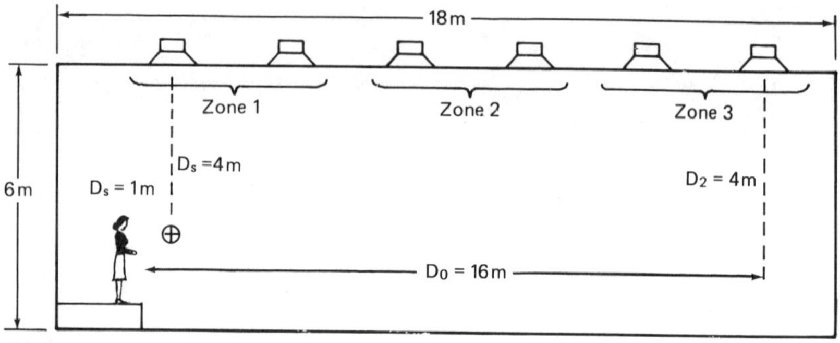

Time delays
Zone 1 = 0
Zone 2 = 9/344 seconds (26msec)
Zone 3 = 15/344 seconds (44msec)

Figure 11-12. A distributed sound reinforcement system.

of 6 dB used in our PAG equations assumes that the system will be equalized. Without equalization, a safety factor of 10 dB might be more realistic.

DISTRIBUTED INDOOR REINFORCEMENT SYSTEMS

Many large banquet and meeting rooms have low ceilings in relation to their length and width, and distributed systems are almost always specified for such rooms. Figure 11-12 shows a side view of a typical room and sound system installation of this sort. It is obvious that a single-point reinforcement system would not work well in this room; for adequate sound levels at the rear of the room, the level would be uncomfortably loud toward the front of the room. Acoustical gain for a single-point system might be difficult to attain, inasmuch as the normally dry acoustics of these rooms would not permit the microphone to be in the reverberant field of the loudspeaker.

Such systems as these are difficult to analyze due to the large number of loudspeakers, but the values of D_1 and D_2, as shown in the figure, allow a general gain calculation to be made. In this case, Equations (5) and (6) would be used.

While personal microphones would ensure the highest gain, the nuisance of accommodating many talkers usually indicates a microphone on a gooseneck adaptor which can be adjusted for each talker. Additionally, acoustical gain may be increase by slightly turning down the loudspeakers nearest the front of the room. The use of time delay in appropriate zones adds to a feeling of naturalness in such systems by helping to localize the sound source toward the front of the room, even for distant listeners.

On an entirely different scale, a distributed system is always called for in large churches and cathedrals. The high reverberation time requires that listeners not be located at a direct-to-reverberant ratio any greater than about 6 dB, and a multiplicity of column loudspeakers on the side walls or piers aimed at the audience area will ensure this. These systems should always be zoned with appropriate time delay, and a single point source up front in the neighborhood of the pulpit should be employed to add focus to the system.

In the operation of the system shown in Figure 11-13, the sound from loudspeaker *A* reaches all listeners first, due to the various time delay zones. The distributed loudspeakers will provide the requisite loudness and intelligibility, but the listener will always localize the sound source at the pulpit, even though the actual level of loudspeaker *A* is relatively low.

In a system of this sort, gain is rarely a problem, and flush-mounted microphones are recommended. A multiplicity of microphones may be used in the chancel area with appropriate control by an automatic microphone mixer.

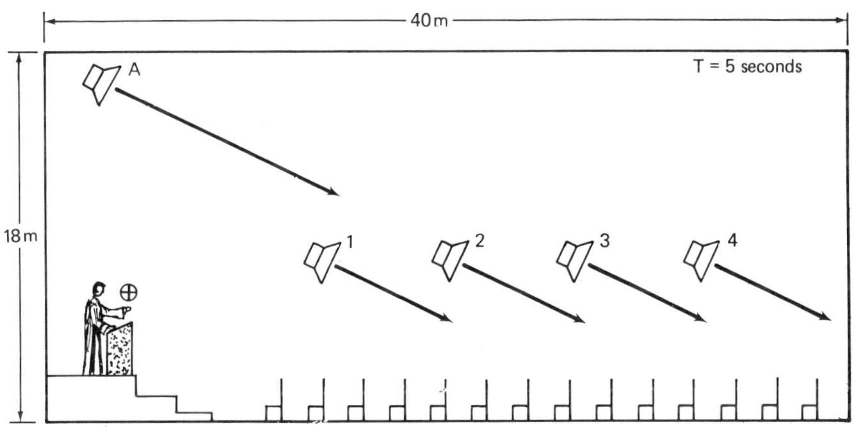

Time delays
Zone 1 = 15/344 seconds (44 msec)
Zone 2 = 20/344 seconds (58 msec)
Zone 3 = 25/344 seconds (73 msec)
Zone 4 = 30/344 seconds (87 msec)

Figure 11-13. A distributed sound reinforcement system in a large church.

SOUND REINFORCEMENT IN THE THEATER

The legitimate theater is usually a fairly intimate experience, and sound reinforcement has not normally been a requirement in the smaller houses. In the larger houses, a slight degree of reinforcement may be helpful, especially for those spectators seated under balcony soffits. A typical theater system is shown in Figure 11-14.

Note the relatively long working distance between the footlight microphones and the action on stage. Accordingly, the potential gain of such a system is relatively low—but then, not much gain is needed. The role of the sound system in this environment is not so much to increase speech intelligibility but rather to restore naturalness and realism to speech through judicious equalizauon. While the proscenium loudspeakers must operate at a fairly low level, it is evident that balcony soffit loudspeakers may allow for considerably greater acoustic gain in that part of the house which they cover. Multi-channel operation is an important part of such a system. At least three stereo channels should be used, and many permanent systems in larger houses have been designed for five-channel operation.

It is customary to operate such a system as this from a console located in the audience area so that the sound mixer can monitor directly what is taking place in the house. An alternate method of monitoring places the mixer at a remote point, where he listens via headphones to a binaural signal picked up by a pair of microphones, as shown in Figure 11-15.

In such a system as this, the mixer wears binaural headphones adjusted so that the level produced by the headphones is the same as the level existing at the microphones. This method of monitoring requires a good bit of experience on the part of the operator. Monitoring by "remote control" may not be a natural thing for many mixers who are used to the intimacy of a live performance.

For musical productions, there would be a pit orchestra located just in front of the stage. The more robust singers would not require microphones, but those of weaker voice would require the complete freedom afforded by wireless microphones. The operator would assign the direction of these signals by means of panpots as the actors moved about on stage.

A general characteristic of multi-channel systems in the theater is that they tend to slightly increase the perceived reverberation in the house. This is caused by the multiplicity of paths available for electro-acoustical feedback and the relatively long time delays afforded by those paths. Properly controlled by the mixer, the effect can be used to good advantage in many otherwise acoustically dry houses.

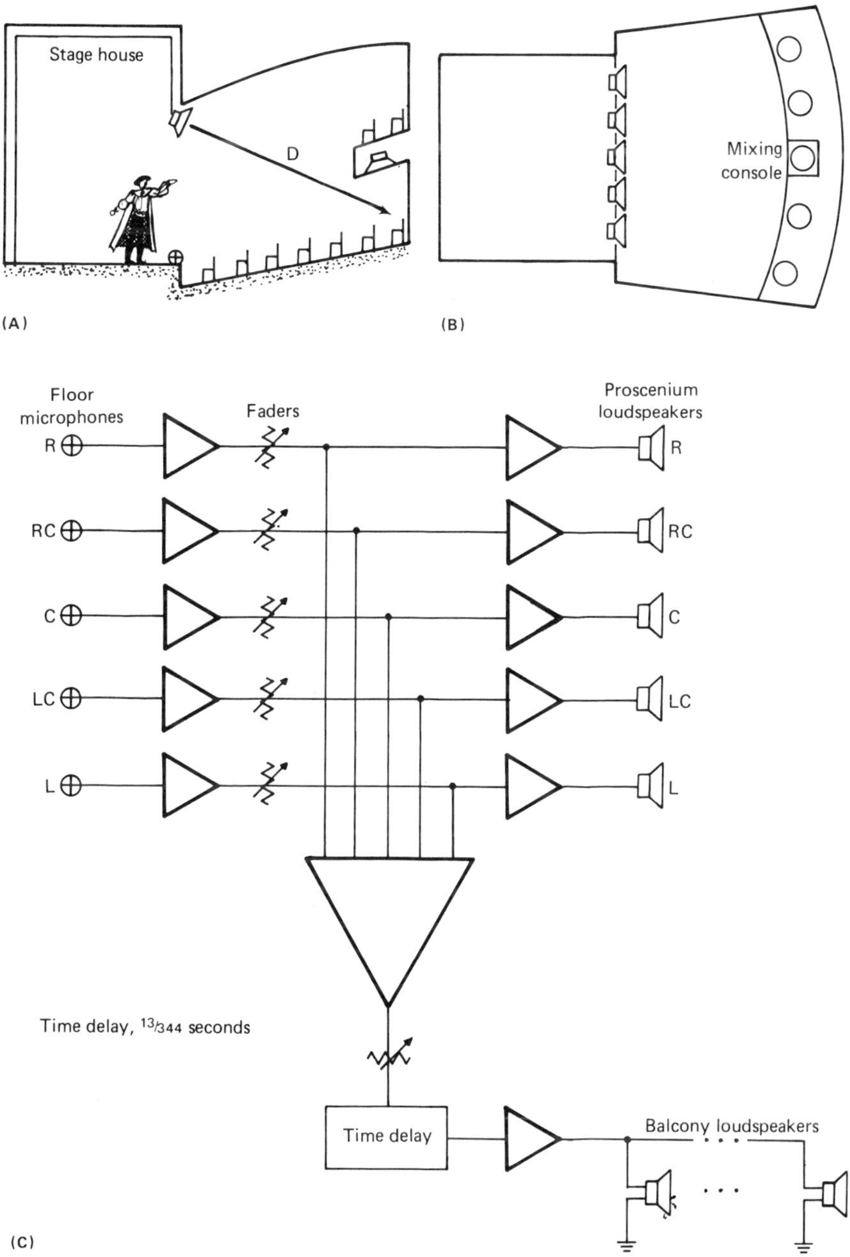

Figure 11-14. A theater sound reinforcement system. **(A)** side view; **(B)** plan view; **(C)** simplified system diagram.

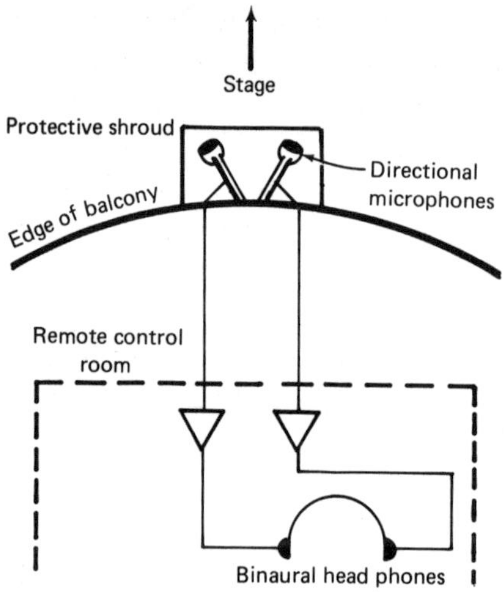

Figure 11-15. Remote monitoring of system via headphones.

OUTDOOR MUSIC REINFORCEMENT

In this section we will discuss two kinds of outdoor music reinforcement, classical and pop-rock. Classical music in its normal indoor environment does not call for any kind of sound reinforcement. Even a single vocalist performing with full orchestra can be heard properly if all musical details are in place. Transporting the orchestra out of doors usually is done in the summertime, in order to attract larger audiences to programs of fairly broad appeal. The main purpose of sound reinforcement in such an application is to raise the level of the orchestra up to a point where it can be enjoyed by those patrons who are located in the farthest seats. The level will rarely be anywhere near what it is in the concert hall, for obvious reasons.

A basic requirement for orchestral performance in the outdoors is a shell around the orchestra. The natural focusing and blending afforded by the shell will usually result in a pleasing balance of sound at sufficient level for those patrons who are close-in. These close-in patrons will rarely be the beneficiaries of any degree of sound reinforcement because they will be well off the axis of the reinforcement loudspeakers.

Another function provided by the shell is that it facilitates picking up the ensemble with a *minimum* for microphones. As a rule, outdoor reinforcement of classical music tends to be "over-produced and under-powered." There is a tendency to use too many microphones and too few of the right kind of loudspeakers. Figure 11-16 shows a typical moderately-sized outdoor orchestra shell. A shell this size would be adequate for up to 4000 listeners and a total "throw" of about 100 meters. Two microphone approaches are recommended; either a coincident or quasi-coincident pair flown above and slightly forward of the orchestra, or an array of three flush-mounted microphones placed close to the edge of the stage. This last technique will work only when the stage is deep and the front of the orchestra is sufficiently set back so as not to dominate in the pick-up.

A coincident stereo pair should bc fcd to a stereo loudspeaker array located *above* the shell, while a left-center-right array or flush-

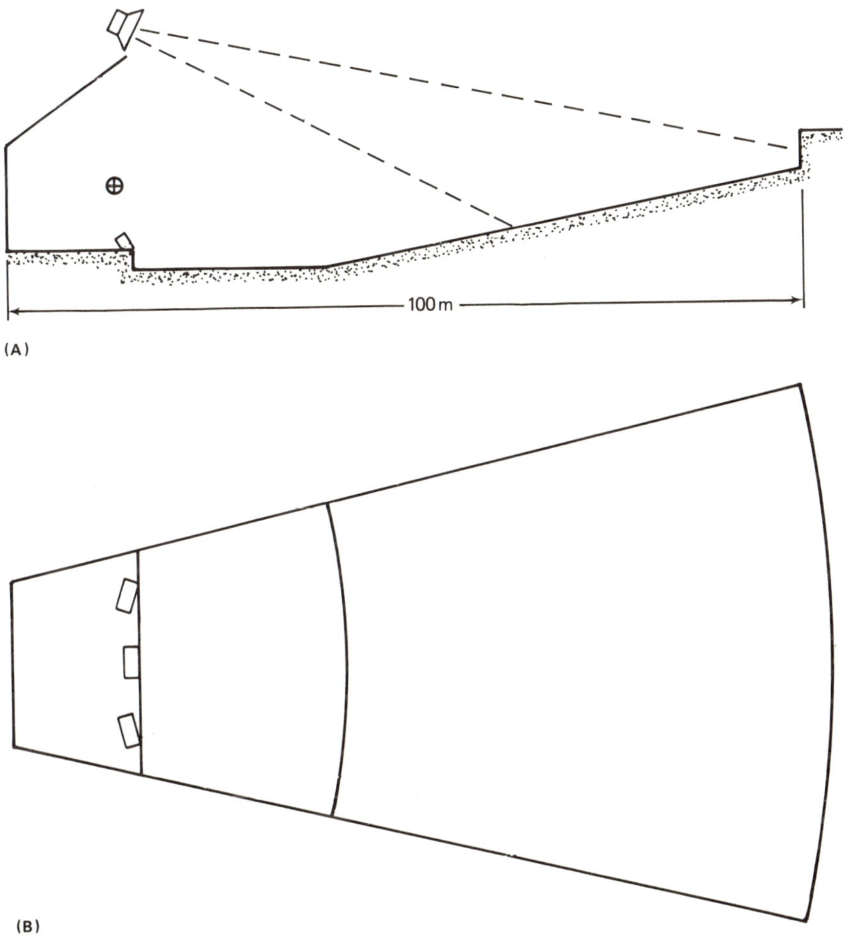

(A)

(B)

Figure 11-16. A medium-sized outdoor theater. **(A)** side view; **(B)** plan view.

mounted microphones calls for a three-channel loudspeaker array. However, by normal splitting of the center microphone's output, a two-channel loudspeaker array can be used. Often the importance of stereo in the reproduction of sound outdoors is not fully appreciated. With mono reproduction, most listeners sense the sound as flat, dull and simply not loud enough. With stereo, most of these objections seem to go away; the sound seems warmer, more natural and more like what one experiences in the concert hall. A reasonable stereo effect may be perceived down to a subtended angle of 20 degrees, so that even the most remote listeners can appreciate the effect. The reason for locating the loudspeakers above the shell is to gain the extra benefit of increased D_1 as well as to ensure greater reach to the rear seats.

In terms of musical dynamics, the problem of peaks not being loud enough is not nearly as great as low-level passages being too soft. Few outdoor environments are really quiet enough to allow an overall dynamic range approaching that of the concert hall. Manual adjustment of system gain is *not* the way to solve this problem; the only remedy is to make sure that the loudspeaker and amplifier resources are up to the job at hand so that the overall range of dynamics can be held as high as possible.

Moving on to the requirements of even larger outdoor systems, let us examine the environment shown in Figure 11-17. There are not many permanent outdoor locations as large as the one shown here. Most are located on a natural sloping depression in hilly terrain: the Hollywood Bowl in Los Angeles is a typical example. Most large venues such as this would be used for all kinds of functions, but we will discuss only the classical orchestral case.

Because of the extreme height of the shell, it will be possible to build into the loudspeaker array elements for near, medium and far coverage. Again, multi-channel reproduction is the ideal, but in an environment this large the hardware requirements will be quite expensive. With a large stage, the best microphone option would be for flush mounting at the forward part of the stage where a natural balance exists through the multiplicity of reflections from all sections of the orchestra.

When sound is heard over distances as great as 200 meters, some strange things happen. Temperature gradients (differences in temperature between the ground and the air immediately above) can affect the long-distance transmission of sound as shown in Figure 11-18. Nothing can be done about the effect when a performance begins

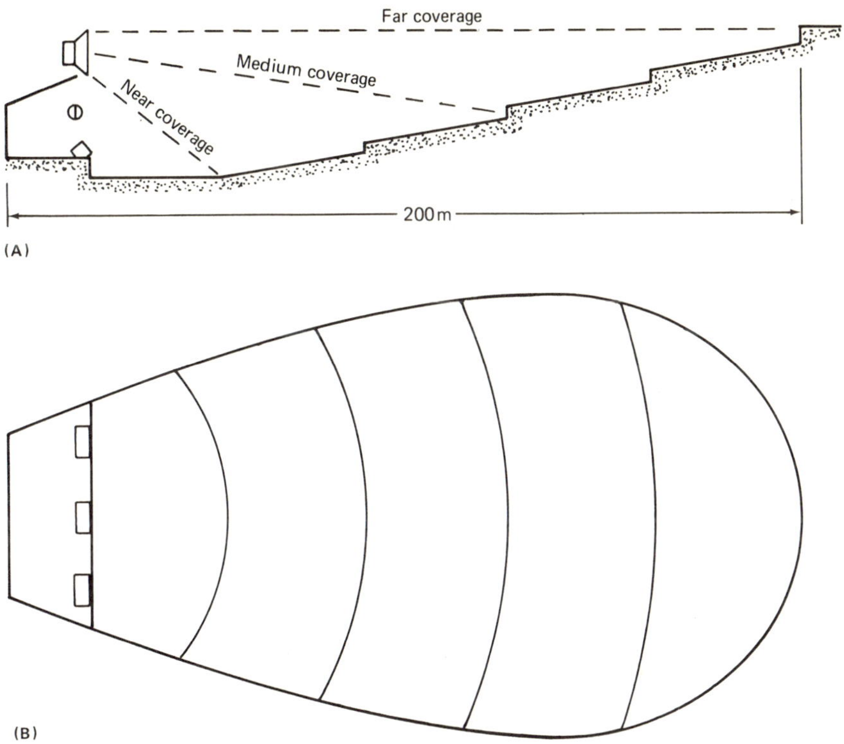

Figure 11-17. A larger outdoor theater. **(A)** side view; **(B)** plan view.

shortly after sunset, when the ground may be still warm and the air above cool, but it certainly calls for an awareness on the part of the system designer that these effects may upset subtle balances established under other conditions.

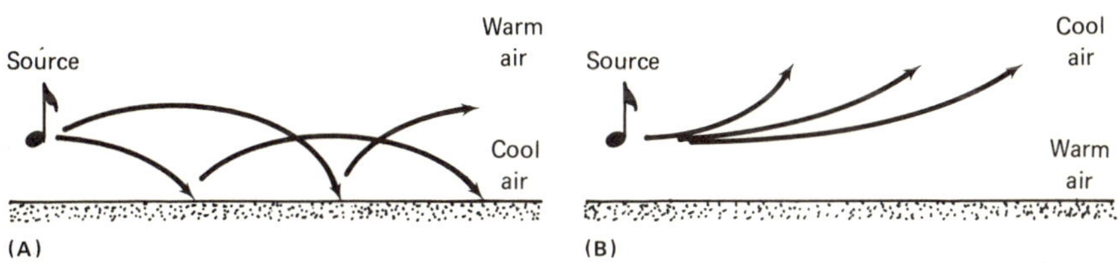

Figure 11-18. The effects of temperature gradients on sound transmission. **(A)** cooler air close to ground; **(B)** warmer air close to ground.

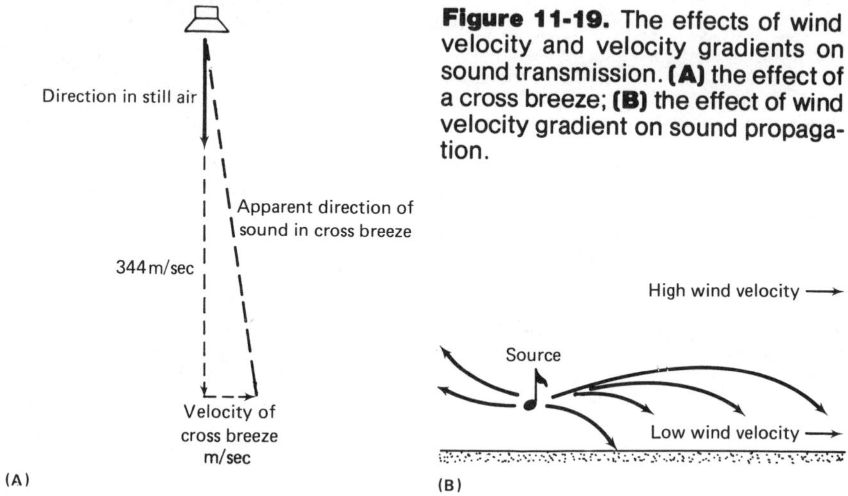

Figure 11-19. The effects of wind velocity and velocity gradients on sound transmission. **(A)** the effect of a cross breeze; **(B)** the effect of wind velocity gradient on sound propagation.

The effects of wind velocity and velocity gradients are shown in Figure 11-19. For a cross wind, the net direction of sound will be the vector sum of the velocity of sound in still air and the velocity of the cross breeze. For winds along the axis connecting the audience and source, it is preferable that the audience be "down wind" from the source, because of the normal velocity gradient resulting in low velocities close to the ground and higher velocities above.

OUTDOOR ROCK REINFORCEMENT

Rock music reinforcement differs from conventional music reinforcement in the following important ways:

1. The use of predominantly electronic instruments with their direct (non-microphone) feeds into the mixing facilities.

2. The use of stage monitoring, or foldback speakers, so that the performers hear themselves on stage at quite high levels. It is the D_1 path between the stage monitors and microphones which is likely to be the cause of feedback in a rock music reinforcement system. As a rule, the reinforcement arrays for the audience area are usually far enough away from the performers as not to be a cause of feedback.

3. The use of extremely low values of D_S. Because of the high levels of reinforced sound on stage, short working distances are required if feedback is to be minimized. Contact microphones can

be used with certain instruments to great effect in that they allow performers almost complete freedom of movement.

In many ways, a rock concert works with musical values previously established in the recording studio. A rock group normally embarks on a concert tour after the success of several record albums, and it is the sound on these discs which the fans will expect. Thus, it is common to find complete mixing and signal processing facilities in conjunction with these concerts.

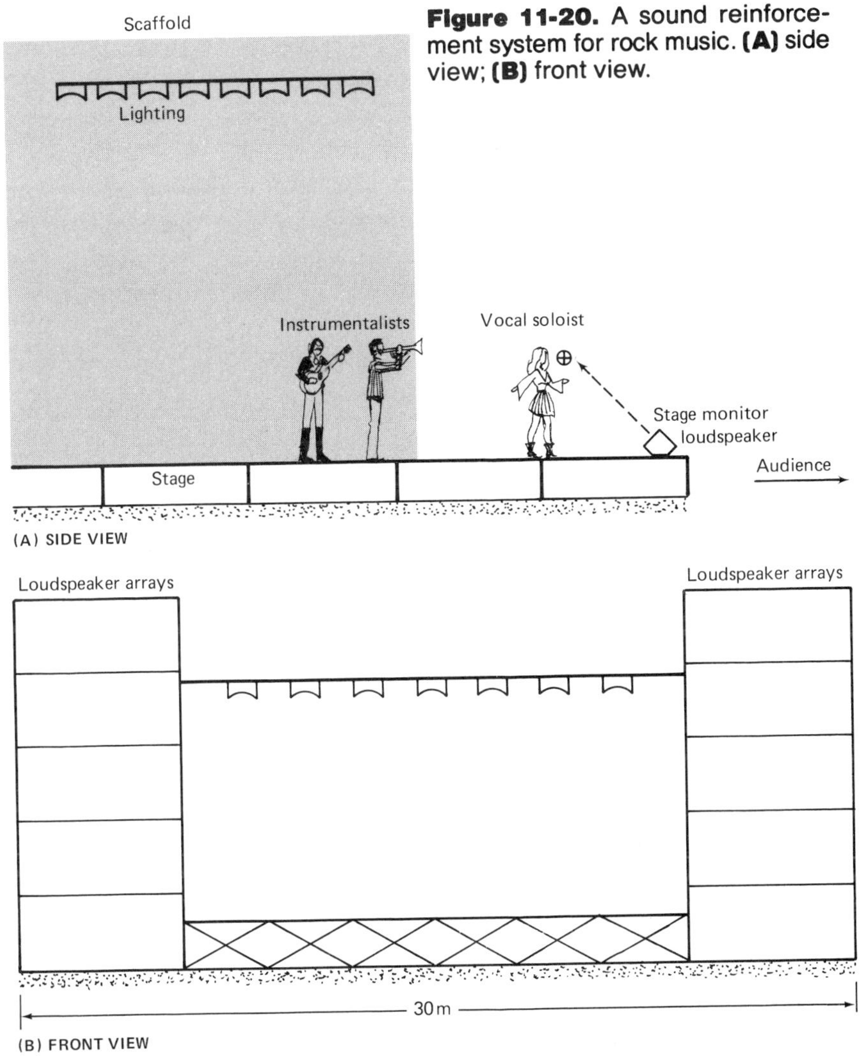

Figure 11-20. A sound reinforcement system for rock music. **(A)** side view; **(B)** front view.

Many spectators are simply awed by the vastness of loudspeakers and amplifiers which are used at large rock concerts. More often than not, these affairs are set up in sports arenas or open fields devoid of any natural reinforcement of sound such as would be provided by a shell. All the reinforcement must come from loudspeakers, and the audiences often number well into the tens of thousands. Since most rock fans like their music in the 110-to-115 dB-SPL range, it is no surprise to see enormous assemblages of low- and high-frequency horn-loaded systems stacked on both sides of the stage. It is not unusual to find up to 10,000 watts of audio power available to drive these large systems. In fact, it could easily take that much power to produce sound pressure levels in the 110-to-115 dB-SPL range at distances up to 100 meters.

Figure 11-20 gives a few of the details of a large rock reinforcement system. Obviously, the hand-held microphone is essential for vocalists, since this is the only way to keep D_S at a minimum. Only the drums would require a typical stereo microphone approach. Here, the normally high acoustical output level of the drums would be the hedge against feedback inasmuch as those microphones would not have to be operated at high gain settings.

CHAPTER XII

Studio Microphone Techniques

INTRODUCTION

The mere mention of studio microphone techniques evokes a wide range of responses from recording and broadcast engineers. Many have evolved their own unique techniques and problem-solving procedures through years of "cut-and-try" experiments, while others have spent their early years assisting and studying more experienced engineers.

Here, we'll offer some "textbook" solutions to common studio problems, with much of the advice coming as a direct application of microphone basic theory. There is always room for other approaches, and the enterprising engineer will keep a good eye and ear out for what others are doing, and for the many techniques which are being developed in studios everywhere.

We will study basic microphone techniques applicable to the several classes of instruments in a variety of studio applications, and we will cover some of the signal processing techniques that go hand in hand with microphone usage. Where appropriate, we will discuss acoustical conditions in the studio. We will not refer to specific microphone brands and models (although we all have our favorites for certain applications); rather, solutions will be presented using specific microphone types. The young engineer is better served by learning to handle microphones in this manner; he will have time later on to develop his own preferences. Finally, we will discuss the role of acoustical separators in the studio.

KEYBOARD INSTRUMENTS

No other instrument poses quite as many pickup problems as does the piano. The sound radiation pattern is complex, and the spectral content may vary considerably with even the slightest re-positioning of the microphone(s). For example, consider the brief musical passage shown in Figure 12-1. Several single-microphone placements are also shown, along with the one-third octave peak-hold spectrum for each position.

CHOICE OF PIANO

The studio trying to make do with anything less than a six-foot grand is making a mistake. Better yet are the seven-foot grands—and of course, a nine-foot grand is pure luxury! The bass registers of smaller instruments tend to produce a wooden sound, no matter how well-regulated they may be.

Conventional wisdom states that pianos improve with age. However, this is not true. A well-built instrument reaches its prime in a comparatively short time, after which it begins a slow period of deterioration.

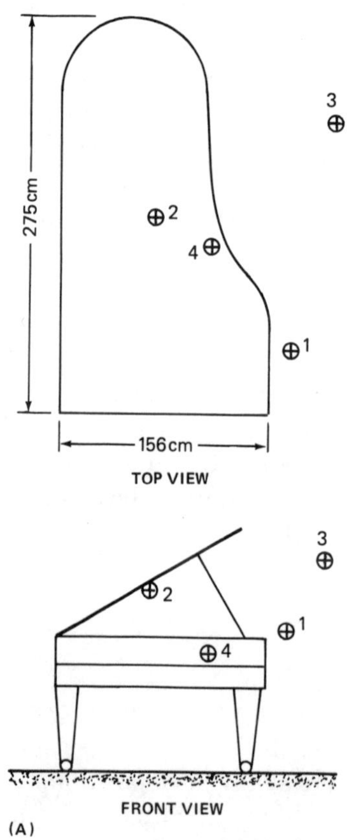

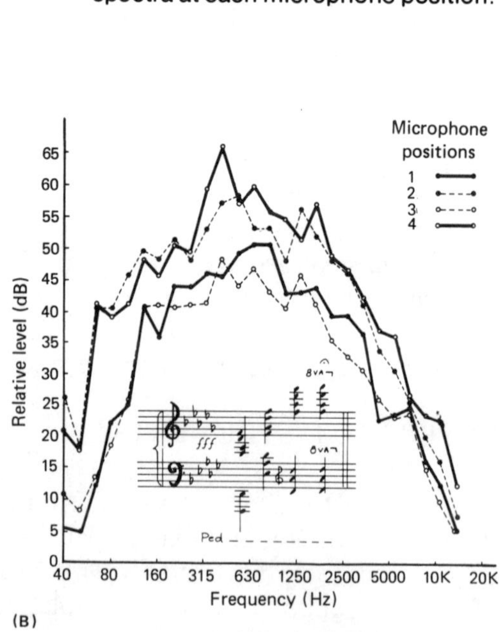

Figure 12-1. Variation of one-third octave peak-hold spectra for various microphone positions in piano pick-up. **(A)** top and front views, showing four microphone placements; **(B)** brief musical passage produces different spectra at each microphone position.

The fact that many newer pianos are not built as well as their predecessors may suggest that the instrument will probably "mature" over the years. But this never happens. A well-regulated new instrument will exhibit even voicing, and just as important, mechanical noises due to the action and dampers will be minimal.

CHOICE OF MICROPHONE PLACEMENT

Some suggested microphone positions are shown in Figure 12-2. The stereo placements are intended for pickup of the solo instrument, and will produce a broad spectrum of sound. A single microphone will generally produce the best mono pickup if it can be located at some distance from the instrument. Alternatively, the stereo pairs may be combined for mono. In doing this, note carefully any apparent

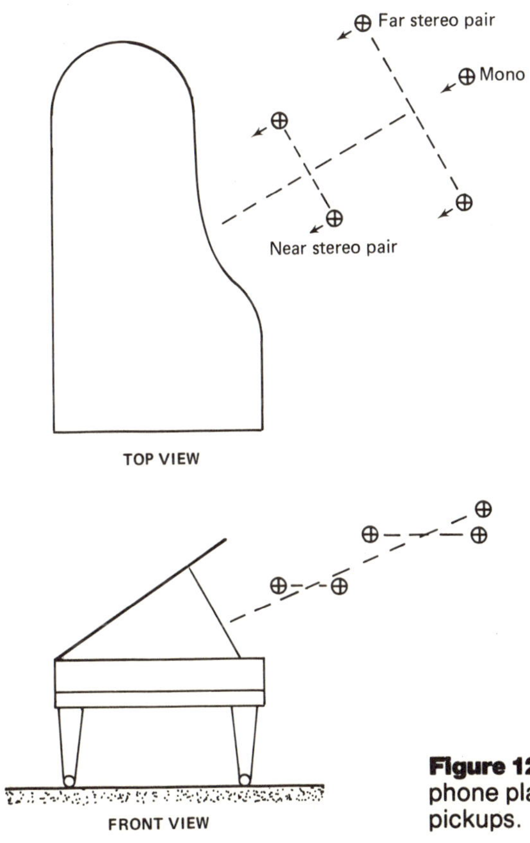

Figure 12-2. Some suggested microphone placements for normal stereo pickups.

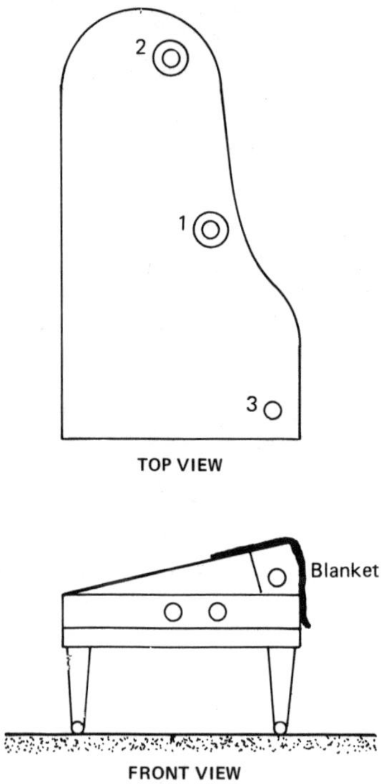

Figure 12-3. A close-up mono pick-up of the piano.

cancellations in the spectrum. An evenly-played chromatic scale through the mid-range will usually identify this.

It should never be necessary to place a microphone under the instrument. Not only is the sound extremely dull, but the thumping noise of the damper action may become obvious.

When the piano is to be picked up in a crowded studio, where the sound level from other instruments is quite high, the techniques shown in Figure 12-3 may be the only ones that will work. A microphone placed in any one of the holes in the cast frame will produce a distinctly unnatural sound, but one which may be quite workable—or even preferable—in pop or rock recording. Omni-directional microphones will be best. If the sound is too "mid-rangy," try adding another microphone at one end of the holes toward the tail of the instrument. If more high end is desired, a microphone may be added above the metal frame close to the top 1½ octaves of strings. With microphone one doing most of the work, microphones two and three should be carefully mixed in, to provide the desired spectral correction.

This is, of course, a *mono* pickup, with the outputs of all three microphones combined.

If even more isolation from loud instruments is needed, a heavy blanket may be draped over the piano lid, with the lid at "half-mast."

The techniques described above should result in a sound well-isolated from anything else in the studio. It will be quite dead, and accordingly may benefit from a slight touch of artifical reverberation. In-line equalizers, especially if they are capable of band-reject action, may help to tone down the brightness of the sound somewhat.

Another important observation of this kind of pickup is the effect of normal dynamic range changes in the playing of the instrument. The piano is capable of an extremely wide dynamic range, and loud passages will always be brighter than softer ones. They may, in fact, become too bright for such close pickup. In that case, the performer should be requested to contain the dynamics within some allowable range determined experimentally. Any additional dynamic control can always be provided by the engineer and producer in the control room. *Never* wait until the time of the session setup to determine the pickup of the piano. It is, after all, one instrument that is always in the studio, or will have been brought in well ahead of the session. Thus, there should be ample time for experimenting. Do not be afraid to tell a pianist to alter playing style, as may be required by the needs of the session. Some players are noisy pedalers, and close microphone placement may pick up too much thumping noise from the dampers. A slight change in technique on the part of the player will usually fix this.

CLASSICAL PIANO

For classical piano, the normal studio environment will not be satisfactory. There are only a handful of really good studios around the world suited for this purpose. The best recording venues are medium-sized live recital halls, or old-fashioned ballrooms. What is most desireable is a room whose reverberation characteristics emphasize the mid- and high-frequency portions of the spectrum. Many halls which are excellent for orchestral recording may simply be too big for piano recording.

A fairly high level of reverberation may be perfectly satisfactory, if it is not too long. Like speech, the piano needs to be well-articulated, and a too-long reverberation time will blur musical detail. It should

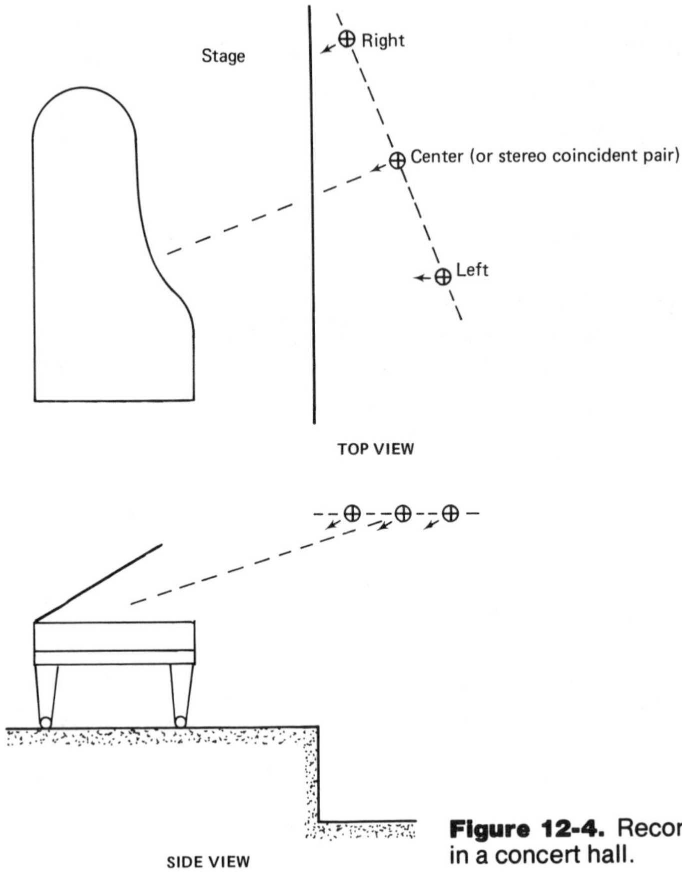

Figure 12-4. Recording the piano in a concert hall.

not exceed about 1.5 seconds. Some details of microphone setup are shown in Figure 12-4.

Many piano recordings suffer from poor focus or imaging in stereo playback. A coincident or quasi-coincident approach will cure this. If a spaced-apart approach is favored, there should *always* be a center microphone to stabilize the image.

Establishing the fore-aft position of microphones is critical. It is very instructive to begin close in and move the microphones outward in *small* steps, perhaps no more than about 30 centimeters at a time. There is often a fairly-narrow range, depending on the room and the instrument, where all musical details seem to fall into place. It is well worth taking the time to find the choice location for the microphone. Especially in classical recording, one cannot stress the importance of a fine instrument enough. Nothing less than a nine-foot instrument should be used, and it should *be in tune.*

HARPSICHORD AND CELESTA

Two other keyboard instruments are notorious for their noisy actions: the harpsichord and the celesta. These instruments are encountered less and less in pop recording, since their specific roles have been pretty much filled by various electronic keyboard instruments.

The harpsichord has the noisiest action of just about any instrument. When a note is released, the plectrum contacts the still-vibrating string before the damper can mute it. In the mid and upper portions of the keyboard, this may not be a problem, but the lower keyboard suffers from various "clunks" and "thunks" of the action. At close quarters, one is aware all the more of the action noise, as well as a high-end "sizzle." Clearly, the harpsichord is an instrument meant for a specific environment; it works best in intimate halls with reverberation times up to say, two seconds. Such environments tend to downplay the raucous high end considerably.

The engineer who is faced with recording a harpsichord as part of a typical pop session has few options. The instrument produces only a moderate level, and is inherently of limited dynamic range. A single omni-directional microphone located close to the raised lid, about one-third down the instrument, will pick up a good balance with minimum action noise. As a rule, its use in pop recording will be relegated to passages where it is exposed and not forced to compete with louder musical resources.

The celesta is rarely encountered these days. Its metal bars produce a full, round sound, often with a good bit of low-frequency action thump. A closely placed omni with a steep 80- or 100-Hz high-pass filter to eliminate the action noise will be the best approach.

WOODWIND (REED) INSTRUMENTS

Woodwind instruments radiate sound in a way which varies with the opening and closing of holes along the sides of the instrument. Figure 12-5A shows the basic problem of interferences due to multiple paths between instrument and microphone. These instruments are relatively limited in their dynamic range, and as a group (the saxophone excepted) they are not loud instruments. As a result, there may be sig-

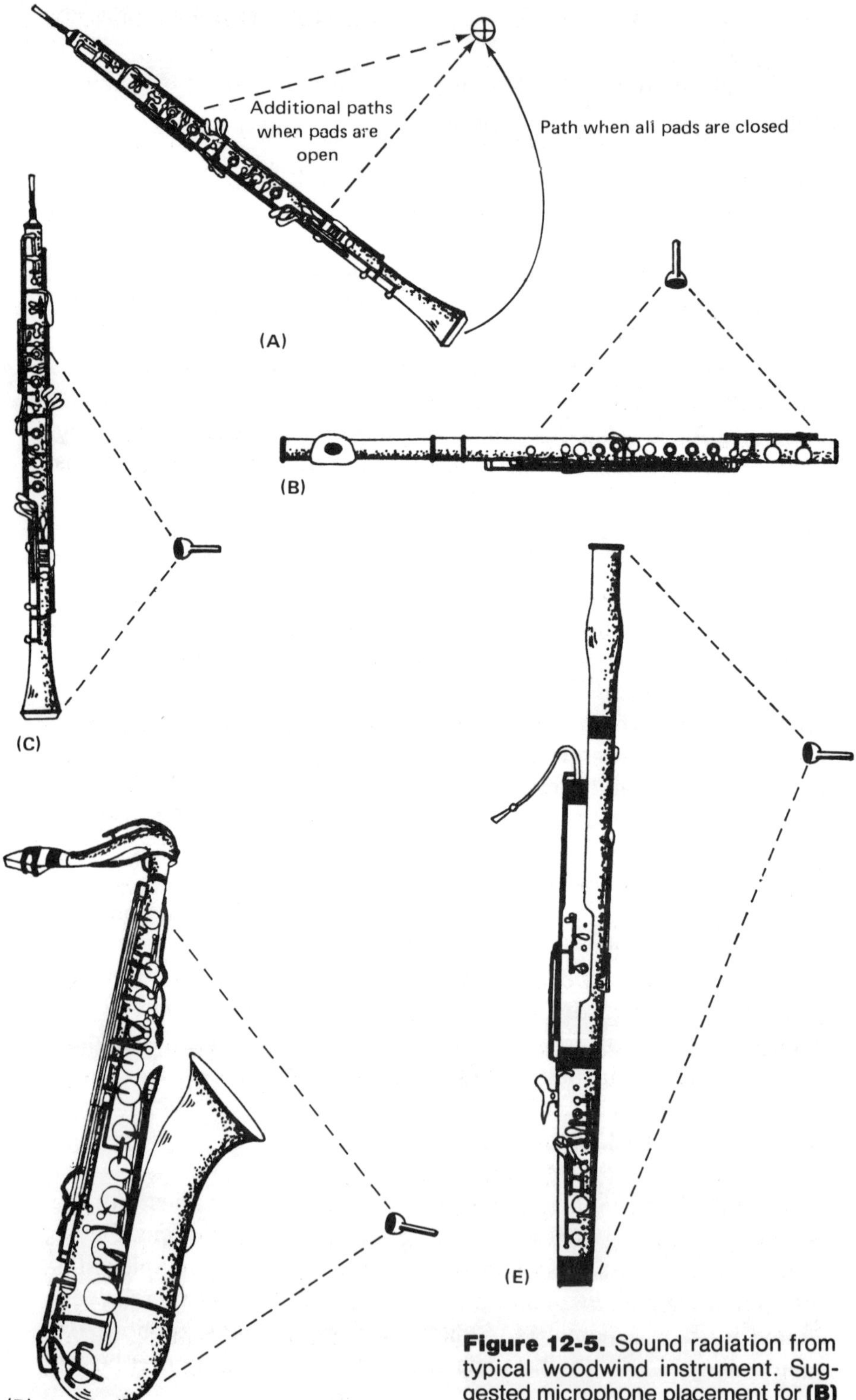

Figure 12-5. Sound radiation from typical woodwind instrument. Suggested microphone placement for **(B)** flute; **(C)** clarinet or oboe; **(D)** saxophone; **(E)** bassoon.

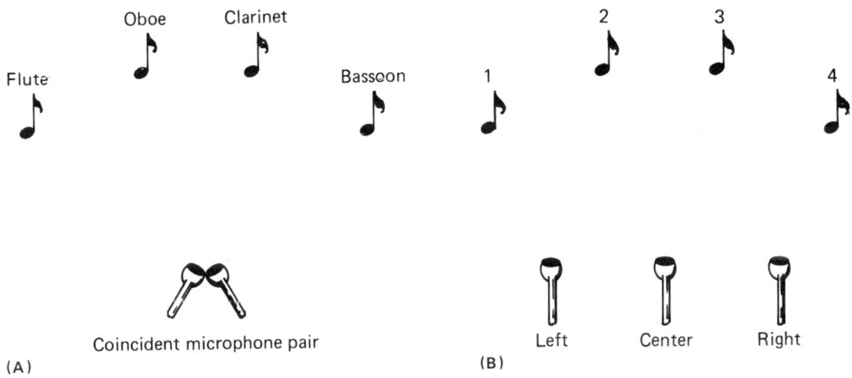

Figure 12-6. Recording woodwind groups. **(A)** no doublings; distances to microphones not to exceed about two meters; **(B)** players doubling; nominal distances between players about 1.5 meters.

nificant leakage of louder studio sounds into their microphones. This potential problem encourages engineers to place microphones even closer to the instruments, resulting in more uneven pickup.

The suggested microphone location shown in Figure 12-5 are general guidelines. As a rule, a directional microphone should be placed toward the bottom one-third of the instrument at a distance such that the array of holes subtends an angle of about 90 degrees at the microphone. Such placement tends to favor lower notes and generally promotes fullness of sound. This suggestion must, of course, be modified as required by local conditions, including distances between players, relative levels of the players and interference levels in the studio.

Woodwinds are often used in choirs. A minimum grouping consists of four: flute, oboe, clarinet and bassoon. In these cases, it is usually advantageous to record them in stereo. Figure 12-6A shows a coincident pair used for picking up a group of four players. Woodwind players often double on two, sometimes three, instruments. Such expanded resources would be found only in conjunction with large studio orchestras, and these doublings may include such instruments as piccolo, alto flute, bass clarinet and contra bassoon, as well as members of the saxophone family. Under these conditions, players will need more space around them, and the simple stereo pickup will have to be modified to a spaced-apart approach, as shown in Figure 12-6B. These arrangements both allow a far more natural pickup of woodwinds than the close pickup approaches shown in

Figure 12-5. The problem is basically one of overall separation from louder instruments in the studio. Obviously, the nature of the music must be compatible with the approach.

An individual woodwind instrument may lend itself to multiple microphone close pickup for mono. For as we have noticed, a single microphone will not pick up a good balance of sound if it is placed too close to the instrument. In some applications, as many as four microphones may be used to cover a single instrument. As with multiple microphone pickup of the piano, the precise locations, and the balancing of these microphones, is empirical. While the performer plays a chromatic scale, listen carefully and make sure all notes are about equal in level. The aim in multiple microphone mono pickup of single instruments should be to attain a spectral balance of sound approaching that which naturally exists in a reverberant performing environment. Such an approach would be quite cumbersome with a group of reed instruments, and it would seem to have its greatest utility in picking up a woodwind soloist performing with a studio orchestra.

BRASS INSTRUMENTS

Brass instruments are capable of a very wide dynamic range, and, with the exception of amplified instruments, they are the loudsest ones which will be encountered in the studio. Their radiation pattern is relatively constant. Most of the energy is radiated along the bell axis, but off-axis positons reveal significant higher overtones being radiated from the body of the instrument. This is particularly true when these instruments are muted. Close microphone placement along the axis of a trumpet will pick up a spectrum quite rich in higher overtones. If the microphone exhibits any degree of high-end rise when used at zero-degree incidence, then trouble may be expected later on during subsequent recording transfer processes. Many engineers prefer older ribbon microphones for close brass pickup because of the generally more mellow high-end response as well as proximity effects boosting of the low end. For single instrument pickup of trumpets or trombones, a microphone would normally be placed about a meter on the axis of the bell. For a soloist with a band, more latitude would be allowed here, and even closer placements may be in order.

A recording engineer needs to be concerned at all times with the *peak factor* of the program at hand. Peak factor is usually expressed in dB and represents the difference in level between peak signal values over a given time period and the average value of the signal over the same period. While percussive transient signals produce the highest peak factors, sustained trumpet notes produce the highest steady-state peak factors. Figure 12-7 shows the waveform of a loud trumpet note with a fundamental of about 400 Hz. This waveform exhibits a peak factor of some 13 dB! What this means is that when the signal is read on a standard metering system at, say, zero VU, actual signal values 13 dB higher are passing through the transmission system. No wonder then that loud trumpets can get out of hand so easily during recording and post-recording operations.

In most popular recording, especially of the "big band jazz" variety, trumpets and trombones make up most of the brass resources. There may be a tuba as well, and, more rarely, two or three French horns. The trumpets and trombones deserve to be treated as a choir and picked up in stereo. Convention dictates that the spread be trumpets on the left and trombones on the right. French horns would fit in the center and the tuba would be placed on the right. A typical pickup is suggested in Figure 12-8.

While trumpet and trombone players normally face the microphones in a forward orientation, the French horn and tuba require different

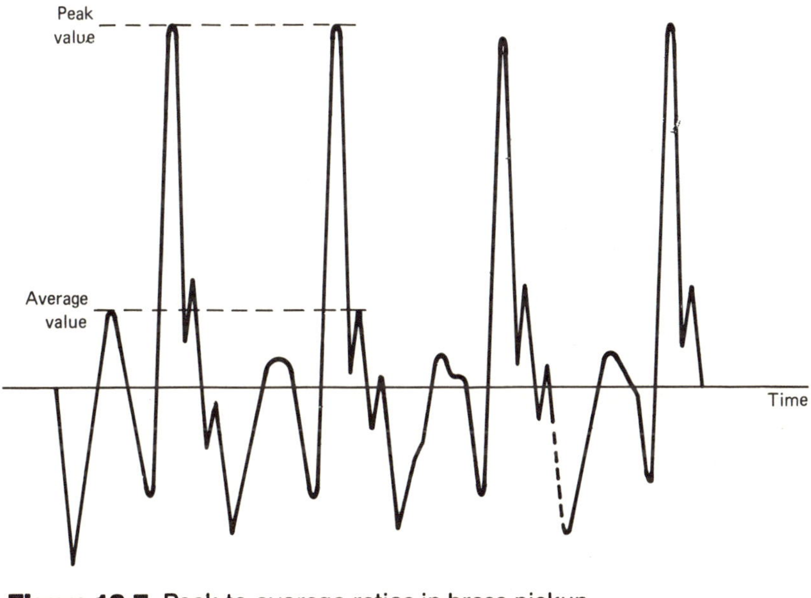

Figure 12-7. Peak-to-average ratios in brass pickup.

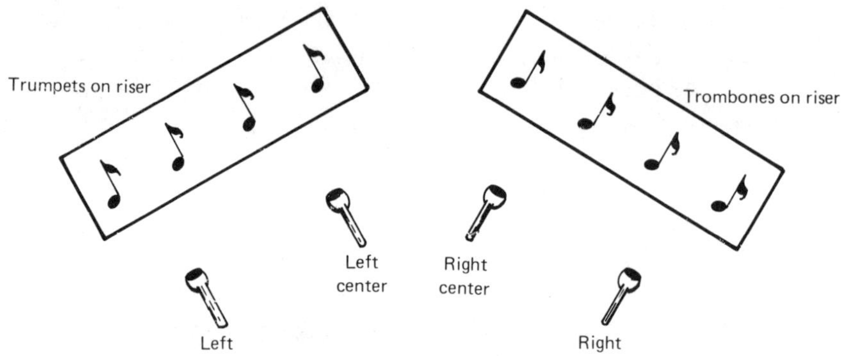

Figure 12-8. Typical brass pickup in the studio.

approaches. The French horn bell is aimed to the right side of the player and slightly to the rear. Many engineers prefer to place the French horn players as far away from the general proceedings as possible and pick them up with a microphone located close to a reflective baffle, as shown in Figure 12-9A. French horns have a way of leaking into every available open microphone. This is by no means objectionable, and it will help to add mellowness to a sound which ordinarily tends to be "buzzy" when picked up at close quarters. The tuba will require its own directional microphone located about a meter above the bell, as shown in Figure 12-9B. These instruments should be panned into the stereo array as required.

STRINGED INSTRUMENTS

Bowed instruments possess a fairly-wide dynamic range, but their overall output capabilities are limited. This, of course, is why they are used in large groups. In most pop recording, the string choir, consisting of violins, violas, cellos and string basses, is used for adding warmth, or for "sweetening" the musical texture in the form of a counter melody or other secondary musical role. Because of this, they are often over-dubbed at a later date, when the studio is free for them alone. A single stringed instrument may be picked up monophonically if that will satisfy the musical requirements. In this case, a single microphone located about 1 or 1.5 meters above the sounding board will suffice. However, where an ensemble is involved, other techniques will be required.

Even though the session budget may demand that you make an attempt, it is difficult to make a handful of strings sound like a large

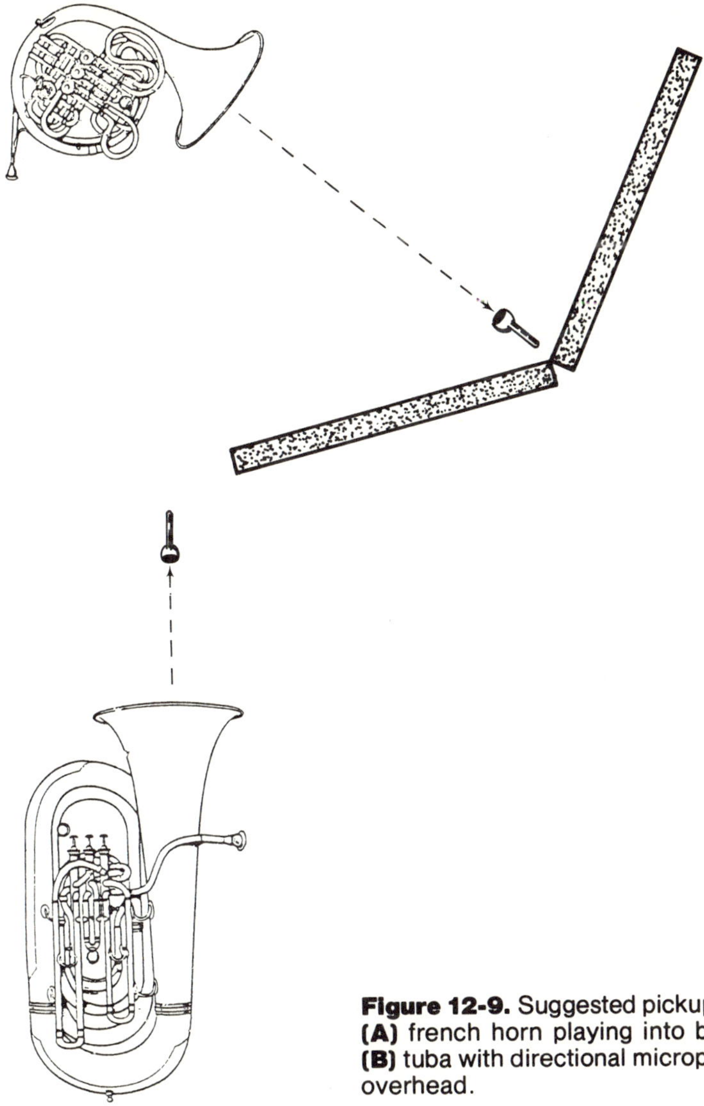

(A)

(B)

Figure 12-9. Suggested pickups for **(A)** french horn playing into baffle; **(B)** tuba with directional microphone overhead.

group. Microphones should not be too close, since that will invariably produce a harsh sound. A natural stereo spread to the string ensemble sound is essential, and multi-mono pickups are usually disastrous. While coincident pickup will produce an integral, well-spaced sound, the particular goals of pop recording will probably be met best through the use of spaced-apart microphones, with sufficient acoustical leakage between them to add the required warmth. In fact, the spaced-apart approach will make the group seem somewhat larger than it actually is.

STUDIO SETUP

Allow sufficient room for the string players. Normally, they are used in pairs, with each pair requiring about 1.5 meters on a side. Microphones should be located overhead, at a distance of about three meters. The number of microphones will depend on the size of the group. A typical group may be made up of six first violins, six second violins, four violas, four cellos and one or two basses. A group of this size may require up to six or seven microphones, each panned across the stereo array as required. An additonal microphone may be needed for the string basses to provide a firm bass line under the control of the mixer and producer.

REVERBERATION

Artificial reverberation is usually a necessity in recording a small group of strings in even a large studio. While some studios can be adjusted to be fairly reflective acoustically, they are usually too small to produce a significant reverberant field so essential to good string sound. Today's better stereo reverberation generators, both digital and analog, are equal to the task. A typical floor plan and signal flow diagram for recording a string section is seen in Figure 10. While this multi-microphone approach may worry some "purist" engineers, it does provide a high degree of flexibility in string recording. Individual microphone inputs are panned to their respective positions in the stereo array. In Figure 10B, the Δ T units represent time delay modules. Those feeding off the stereo mix would be set in the 30-to-50 millisecond range, where their effect would be to simulate early reflections in a concert hall. The delay module in series with the reverberation generator would be set in the 40-to-70 millisecond range, in order to simulate the normal delay in a concert hall before the onset of reverberation.

The various plucked stringed instruments, such as the harp, guitar, and so forth, usually produce a fairly soft sound. The best approach is to use a single directional microphone located perpendicular to the sounding board at a distance just large enough to ensure even pickup from the entire sounding board. If leakage problems require closer

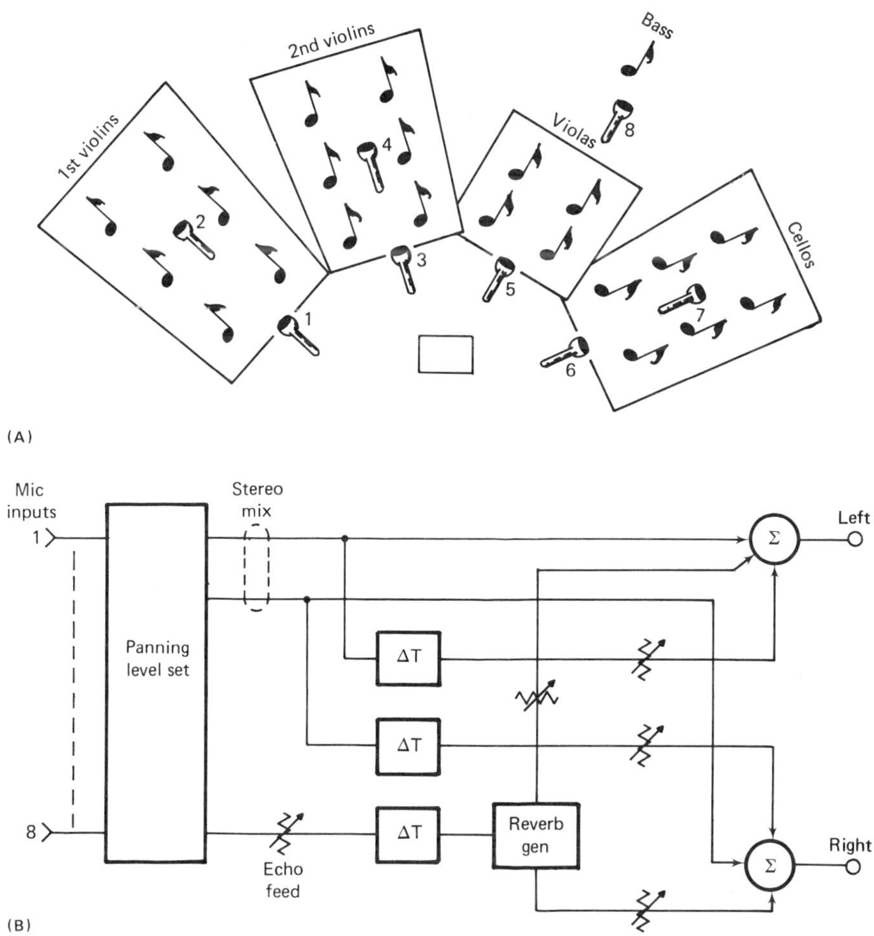

Figure 12-10. Signal processing scheme for studio pickup of a medium-sized string ensemble. **(A)** floor plan; **(B)** the signal flow diagram.

placements, be prepared to roll off a bit of low end in order to retain natural balances. Typical pickup of the harp and guitar are shown in Figure 12-11.

The string bass may require attention. It is not unusual for studio orchestras to have only a single string bass, and in order to pick up sufficient fundamental sound, a single microphone placed on a low stand in front of the instrument will be needed. Its output will have to be panned and balanced into the stereo array carefully.

For jazz recording, the string bass takes on an entirely different role, and it needs to be highlighted as a virtuoso solo instrument. It is

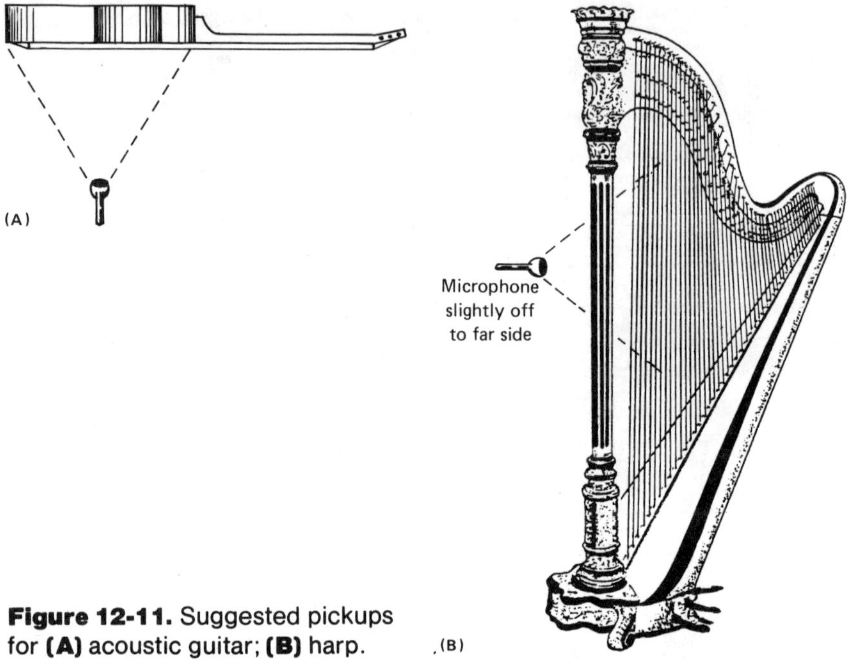

Figure 12-11. Suggested pickups for **(A)** acoustic guitar; **(B)** harp.

almost always played pizzicato, or plucked, in jazz work, and this demands that finger noises and rattling of the strings on the finger board be picked up accurately, rather than suppressed. A single microphone at a height of about one meter, aimed at the front sounding board, at a distance of about 0.5 meters, will insure a good pickup. Be sure to thin out the bottom end a bit if the sound seems too muddy.

ELECTRONIC INSTRUMENTS

While microphones may be placed in front of the loudspeaker enclosure used with a given electronic or amplified instrument, the electrical output from the instrument may be plugged directly into the console. Quite often both are used. A so-called "direct box" is generally used for the electrical feed, and it provides electrical isolation to inhibit hum, buzz and other undesired noises. Even then there may be problems. Under the best of conditions, many electrical pickups are noise-prone, and the direct-in application seems to enhance this all the more. Proper equipment maintenance is called for, and noise gates can often be used to good effect. A typical "direct box" application is shown in Figure 12-12.

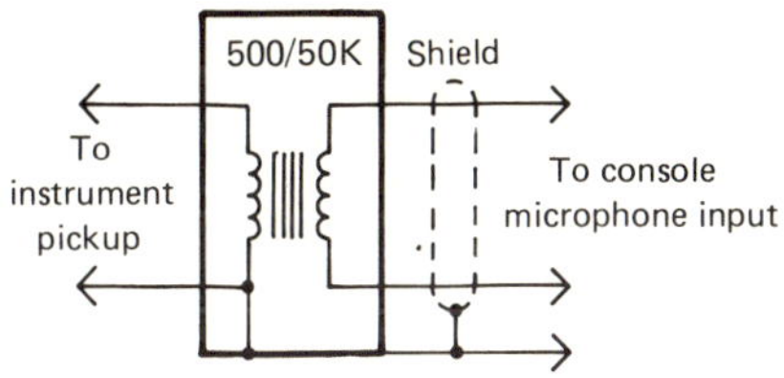

Figure 12-12. A typical "direct box."

The electric bass, whether picked up direct or by microphone is often heavily compressed, as shown in Figure 12 13. If a compressor is used whose time constant is long with respect to the natural decay rate of the instrument, the result will be as shown in Figure 12-13B; the basic quality of sound is maintained, but the level is held at a predetermined point. If the recovery time of the compressor is short

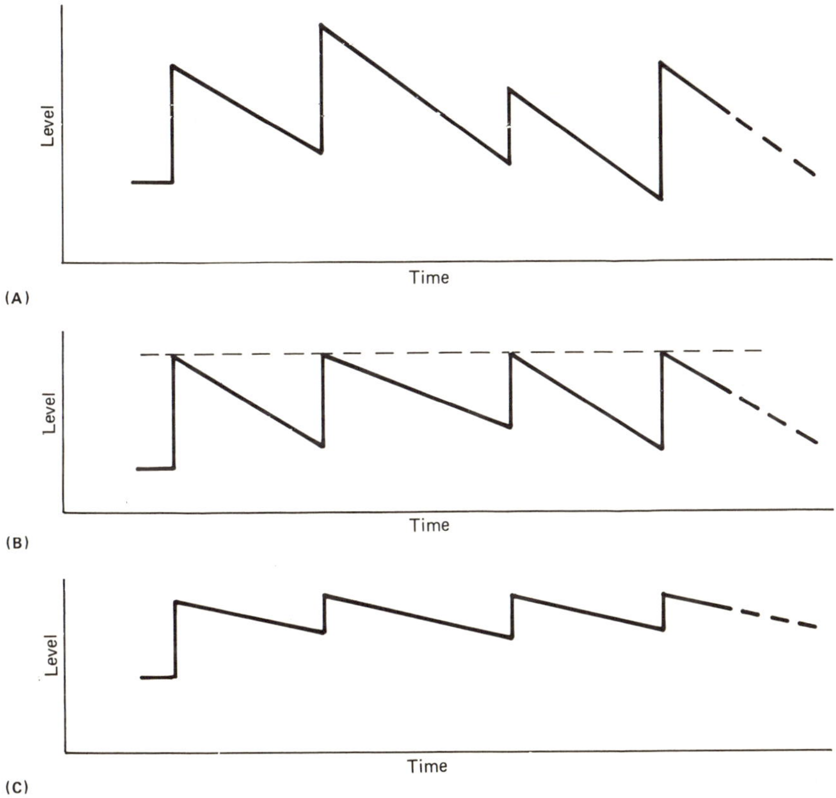

Figure 12-13. Compression of the bass line, showing the bass envelope **(A)** uncompressed; **(B)** compressed, with slow recovery time; **(C)** compressed with fast recovery time.

compared to the natural decay rate of the instrument, the envelope will be modified as shown, and the instrument will take on an organ-like quality.

PERCUSSION INSTRUMENTS

Percussion instruments come in all sizes and shapes. The smallest are the triangle and finger cymbals, and the largest are the bass drum and the gong (tam-tam). Figure 12-14 shows a typical drum kit setup for a rock recording session, with typical microphone locations. In rock recording, the drum set is invariably picked up in stereo. It matters not that the resulting sound, spread as it is from left to right, conveys such a bigger-than-life impression.

The basic components of a drum set are, starting from the bottom, the kick drum, a variety of tom-toms (usually three, tuned low, middle and high), a snare drum, a variety of cymbals, and a "high-hat" set of cymbals. It is impossible for a single microphone to do justice to a drum set, even for a mono pickup. Many closely spaced microphones are necessary if crispness and intimacy of sound are to be maintained.

While these instruments are capable of a wide dynamic range, they

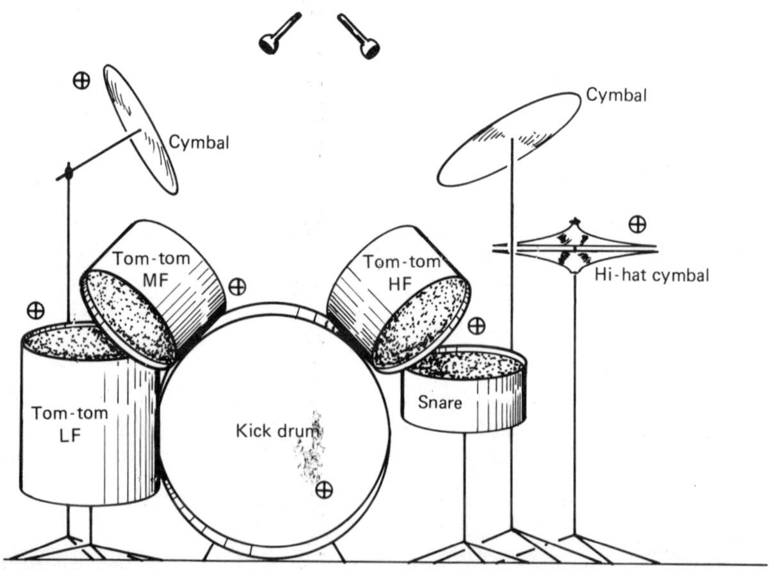

Figure 12-14. Pickup of the standard drum set.

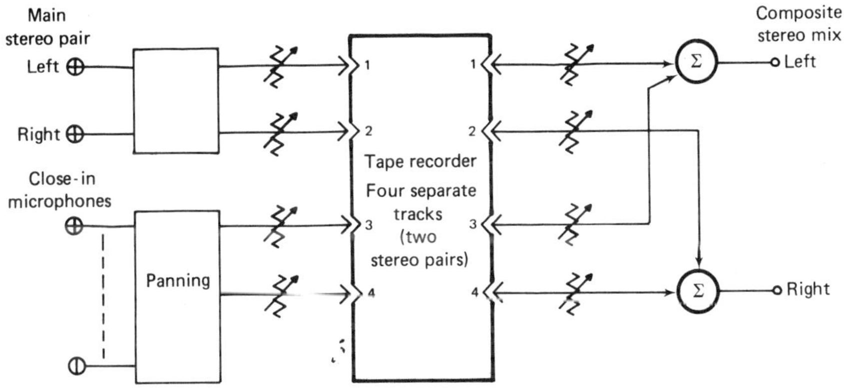

Figure 12-15. Recording the drum set, using two stereo pairs of tracks.

are most often played *loud.* In a typical studio setup, leakage from the drum set into other microphones is a problem. Jazz drummers typically make good musical use of the dynamic range available to them and tend to keep their maximum levels somewhat lower than rock drummers.

Since the drum set is one of the loudest sound sources in the studio, directional microphones will be of little advantage in holding down leakage from other instruments. As a rule, the overall stereo pickup may be relegated to a pair of cardioids, while omnis should be used for the close-in pickup on individual elements. The microphones picking up the individual components of the drum set would be panned to their approximate positions in the stereo array. They should be placed fairly close to the edges of the tom-toms and snare drum. Those placed close to the cymbals should be located carefully so that they will not be touched by the cymbal as it moves while being played.

It is a confident engineer or producer who will record the drum set on a single stereo pair of tracks. Depending on the number of tracks available, he may relegate the drum set to four or more tracks to allow post-recording mix-down flexibility. A convenient set of assignments is to put the overall stereo pair on two tracks, and use two more tracks for the panned signals. In a sense we have *two* stereo pairs, whose combination can be refined during remixing, as shown in Figure 12-15.

In live performance, one never hears the detail out of a drum set that we expect in a recording. Small resonances or vibrations in the individual instruments, which might go unnoticed in normal performance, may create problems in recording. The offenders here are usually buzzes, rattles and metallic ringing sounds. With care, these can usually be remedied through the application of tape, cloth, felt or almost anything that can be used to damp out an unwanted noise. It takes time to do this, and, as with the piano, attention to such details prior to the recording session will be appreciated by everyone. A good studio drummer will normally be well aware of these problems.

There are a variety of signal processing techniques useful in modifying drum sounds. Equalization should be used sparingly, since there is such a wide range of sounds to be obtained merely through attention to microphone placement. *Never* should equalization be resorted to in order to correct microphone type or location. Noise gates on each microphone can be used to tighten up the sound. Careful attention to threshold and recovery time settings is required. Artificial reverberation should be used sparingly, if at all. Even small amounts have a way of muddying up the musical texture.

Some studios use special housing to contain the normally high levels of a drum set. Such "drum cages" are usually enclosed (to a height of about one meter) and have a roof on top. The inner linings are as absorptive as can be made. Drummers do not generally like the feeling of playing in such surroundings, but there is no question that the enclosure does cut down on the ambient level in the studio due to the drums. Details of a typical drum cage are shown in Figure 12-16.

The thicker the wall and ceiling sections of the cage are made, the better the low-frequency absorption. Many newer studios include such enclosures built into a studio wall for best acoustical, as well as visual, effect.

Other drum-like percussion instruments encountered separately in the studio include the bongo drum, tom-tom family and timpani. Those drums which are of cylindrical shape with skin, or heads, on both sides, behave as *dipoles* at low frequencies. Their basic response is as shown in Figure 12-17; one head exhibits positive displacement, while the other exhibits negative displacement. Figure 12-17A shows the proper way to locate a closely-placed microphone, which should be placed on one side or the other and never halfway, as also shown in the figure. The latter arrangement produces conflicting phasors at the microphone, and the result will be reduced fundamental output.

In normal orchestral use, the bass drum is oriented as shown in Figure 12-17C. While most convenient for the performer, the dipole action may attenuate the fundamental, giving rise to a weak low end.

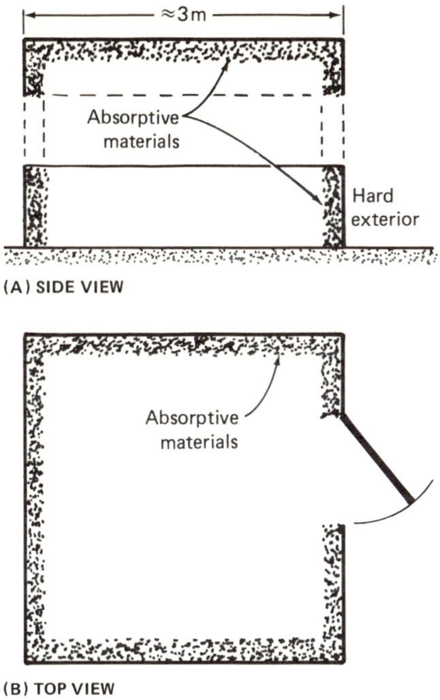

Figure 12-16. A typical drum booth. **(A)** side view; **(B)** top view.

The orientation of the instrument, as shown in Figure 12-17D, may correct this problem. Again, do not be afraid to request that performers change their routines and habits to accommodate the needs of a recording session.

If the microphone is placed next to the head which is struck, the high-frequency content of the pickup will be greater than if the

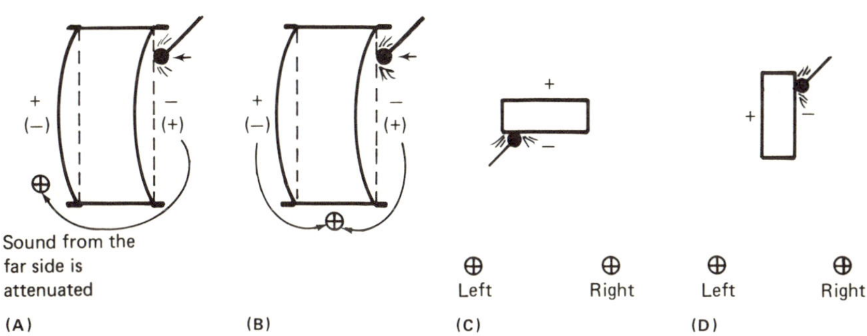

Figure 12-17. The dipole response of a drum, showing proper **(A)** and improper **(B)** close pickup, and proper **(C)** and improper **(D)** placement of bass drum for stereo pickup.

microphone is placed close to the back side. The problem here is that rapid movements of the heads may create slight wind noises. Where a player cannot contain his enthusiasm, windscreens may be helpful.

The timpani are always used in multiples, never less than two and often as many as four or five. An overall stereo pickup may be useful for many studio orchestral sessions. Normally, that stereo pair of microphones would not be spread across the entire stereo stage, as in the case of the drum kit, but rather would be panned to occupy a narrower angle off to one side. The initial impact of the timpani is sometimes lost. It may be restored by carefully mixing in individual omnis located within a few centimeters of the heads. A single omni can easily handle a pair of timpani. Figure 12-18 shows some details of timpani pickup. Be prepared to thin out the texture of these closely picked up instruments with a subtle bass cut if the sound seems too thick.

Metal-to-metal transients, such as produced by triangles and tambourines, produce very large peak factors and can result in tape saturation if special pains are not taken. While these instruments often require close placement, do not move in any closer than necessary. Do not use overly bright microphones; older ribbons often work best. *Never* boost the high end, and be aware of electrical overload in the input stages if fairly high output microphones are to be used. The same observations apply to the glockenspiel or orchestra bells.

The vibraphone and chimes are best picked up by directional microphones located far enough away to subtend an angle of 90 degrees. For the vibraphone, the microphone should be placed over the middle of the bars. The same applies to the marimba and xylophone. For the chimes, the microphone is best placed behind the set, slightly overhead and aimed downward toward the upper half of the tubes.

VOCAL PICKUP

Some vocalists grow up with a microphone in their hand and are uncomfortable without it. Proper studio vocal technique calls for the microphone to be on a sturdy boom hanging in front of and slightly above the performer. The most significant problems faced in recording a vocalist arise from both the wide dynamic range of the voice, as well as the tendency of the performer to move closer to and farther away from the microphone. Normally, the microphone (uni-directional,

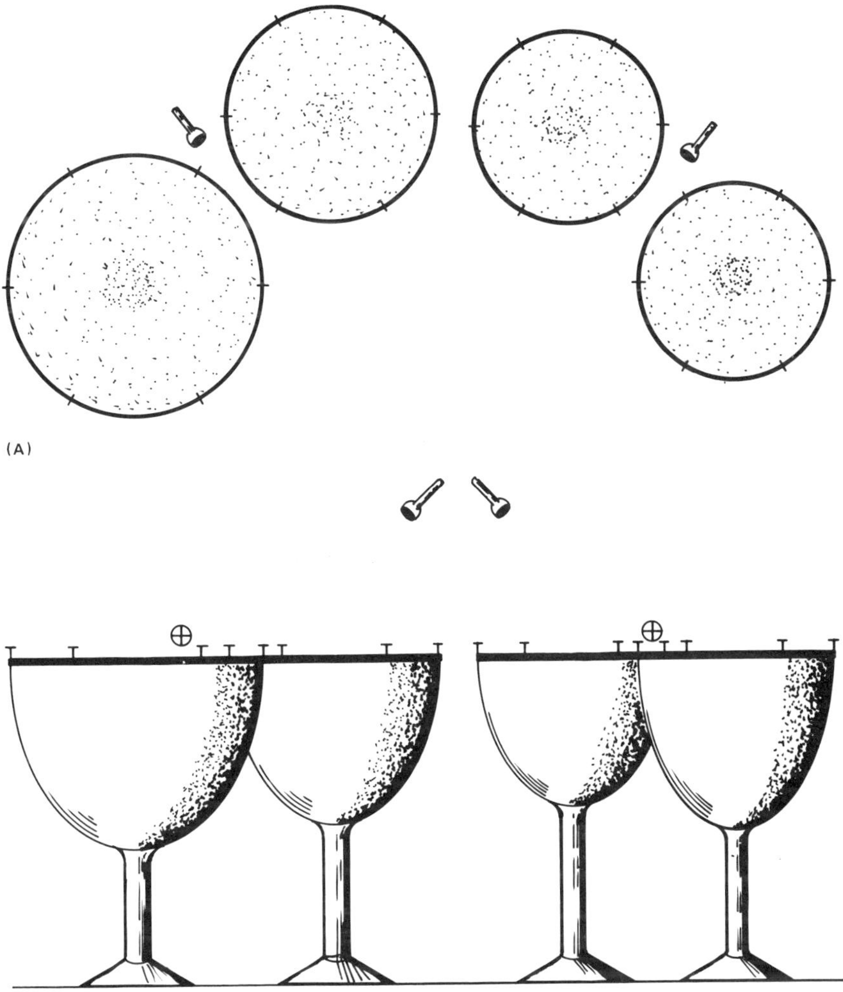

(A)

(B)

Figure 12-18. Timpani pickup. **(A)** top view; **(B)** side view.

or in special instances bi-direction) would be about 10 to 20 centimeters from the vocalist's lips. Locating the microphone above the vocalist allows good visibility of other studio activities as well as of any notes or copy on a music rack.

The problem of level variation is best handled through the use of a compressor; one that will handle a large input range. The engineer will have enough to do without riding gain manually on a wayward vocalist. It may come as a surprise to many readers to realize that pop or rock vocalists are rarely recorded onto disc with anything resembling their actual dynamic range at the microphone. 12 to 15 dB of compression is not uncommon. There is still room however for subtle manipulation of gain *after* the compressor, and this is rightfully the responsibility of the engineer and producer.

Some studios have *vocal booths,* small rooms giving the requisite isolation for vocal work. However, where a vocalist is not being interfered with by loud instruments, or where a vocal line is being added to existing tracks, it is best to record the vocalist in the studio. Vocal booths often are not very good, and they can give rise to a "boxiness" or coloration in the mid to lower registers due to insufficient low-frequency absorption.

Some vocalists prefer standing, while others prefer a tall stool. For recording purposes, the stool is preferred, since it better defines a reference position for the vocalist relative to the microphone.

The vocalist will always need some kind of foldback or cueing, usually via headphones, so that the all-important rhythm tracks can be clearly heard. Headphone monitoring must always be available, and the vocalist should be allowed to specify the particular mix of instrumental tracks for greatest comfort and assurance. Some vocalists will also want to hear their own voices through the cue bus (with or without reverberation). Monitoring levels sometimes run fairly high on headphones, and it is possible that sound from the headphones can cause feedback through leakage into the vocalist's microphone.

Typical relationships are shown in Figure 12-19. Some engineers have experimented with the loudspeaker monitoring scheme shown in the figure. This approach *may* work in a vocal booth where two well-matched small loudspeakers are located equidistant from the microphone. The signal cancellation along the null plane between

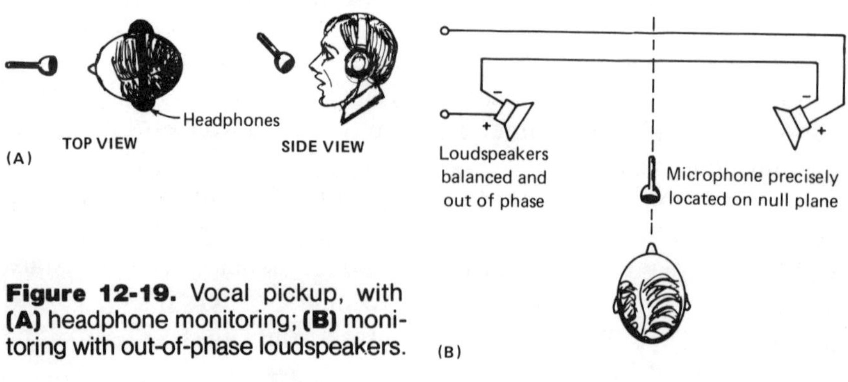

Figure 12-19. Vocal pickup, with **(A)** headphone monitoring; **(B)** monitoring with out-of-phase loudspeakers.

them allows the monitoring level to be raised higher than one might think. However, the vocalist must get used to the curious, often disconcerting, sound of out-of-phase program.

While a good bit of rock vocal recording is done without added reverberation, most pop vocal recording calls for reverberation. The excesses of the sixties are to be avoided. Newer reverberation devices, both analog and digital, offer excellent quality.

Some singers are excessively sibillant and need special treatment before remix activities get underway. Sometimes, a rolled-off microphone will do the job, but more often than not this will take some of the life out of the signal. The best solution is to use a "de-esser," a special limiter responsive only to high frequencies in the program.

In recent years, *out-of-band* generators have become popular. These are devices which, through the controlled generation of distortion, add an extra and desirable top octave where one does not normally exist. They add what no equalizer can add; an equalizer cannot boost what is not present. All things considered, a vocal track may make use of every signal processing trick in the book. (One can easily imagine a vocalist whose wrong kind of highend is attenuated with an equalizer and then right kind of high end is added by means of out-of-band signal generation!)

The speaking voice often needs special attention. Sibillance control is important, as is the basic balance of low, mid and high-frequency energy. Any judgements made by the engineer regarding boosting or cutting should be done over a reliable monitoring system, one that closely approximates the final playback environment. It is also important that any decisions being made regarding equalization be done while monitoring at a level approaching the intended playback level. For example, a speech track intended for low-level playback may require more bass boost than one intended for high-level playback.

Even more than singers, speakers will not fare well with poor acoustic conditions. A marginal vocal booth may work with a singer but might add too much "boxy" coloration to a narrator's voice. The worst offenders by far are the poorly constructed announcing booths with pervade the broadcast industry. These vary from small booths, which add a "honkiness" to the voice, to poorly damped rooms whose own "small-room" reverberant character intrudes. A tasteful amount of high quality artificial reverberation will always enhance a narration track, particularly if it is of a dramatic nature. Small amounts of reverberation may not be noticed as such but will increase the overall spectral energy density and loudness of the signal.

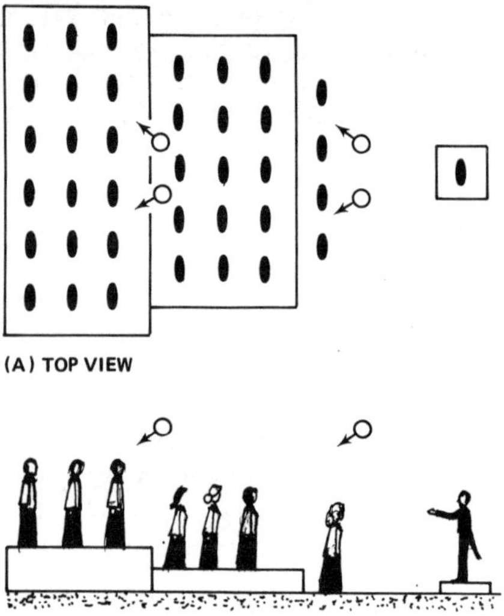

Figure 12-20. Vocal chorus pickup. **(A)** side view; **(B)** plan view.

Like bowed string instruments, vocal choruses should always be recorded in stereo. As with strings, there are usually not enough of them, and there is always the desire to have them sound like more. The approach shown in Figure 12-20 will work for a vocal group performing either in sections or dispersed. As with strings, the use of spaced-apart microphones will make the group sound larger. Artificial reverberation is always called for. Solo passages sung by members of the chorus may or may not need to be highlighted with close-in microphones, depending on the nature of the music.

ACOUSTICAL SEPARATORS IN THE STUDIO

Acoustical separators are screen-like structures which can be rolled around a studio as required to provide isolation between instruments and instrumental sections. They vary considerably in size, and the larger ones are more effective than the smaller ones. Essentially, separators block the direct sound and primary reflections between a microphone and unwanted instruments. Figure 12-21 shows the measured response between a loudspeaker and a microphone with and without a large separator. In this case, we are looking at the normalized (flat) response of the loudspeaker as picked up by the microphone and noting the effect the baffle produces. Considerable mid-range and high-frequency attenuation is provided by such baffles. The

Figure 12-21. Acoustic separators in the studio. **(A)** top view; **(B)** side view; **(C)** response with and without separator; **(D)** top view of studio.

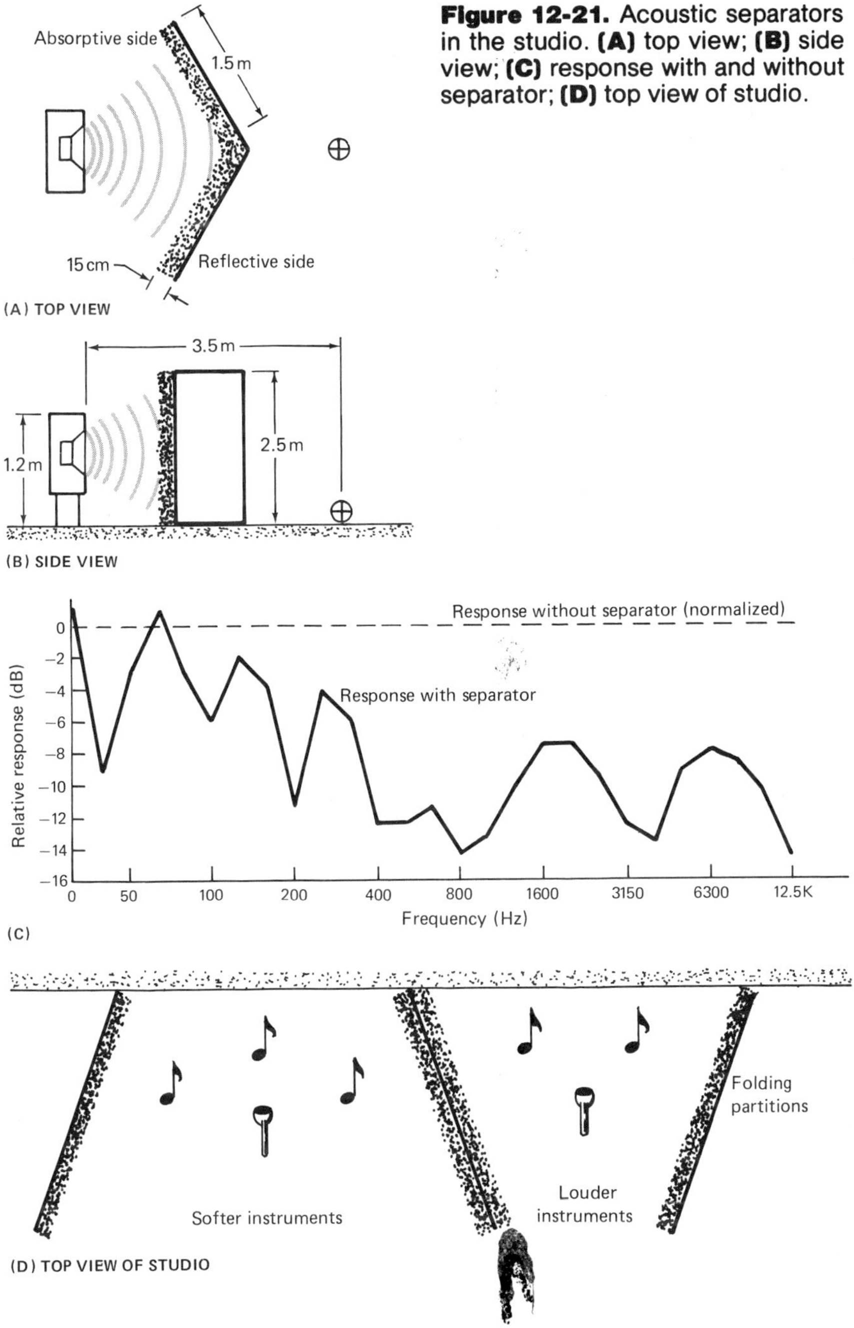

extent of the effect will depend on the nature of the studio. In general, the deader the studio, the more effective isolators will be.

They come in all sizes and shapes. Many are of moderate height, a meter or so, and they have the advantage of allowing players to have full visual coverage of the studio.

Most separators are absorptive on both sides; others are absorptive on one side and reflective on the other. The usage is obvious; the reflective side would be used toward softer instruments while the absorptive side would face the louder instruments.

Separators such as we have described here are normally constructed by the studio as required. Many large studios accomplish similar results by installing folding wall sections, as shown in Figure 12-21D, so that loud and soft resources may be arrayed around the studio for best effect.

CHAPTER XIII

Electrical Interface Considerations

INTRODUCTION

In this chapter we will discuss the many aspects of interfacing the microphone with the input stages of the audio control system. The topics will include: line losses, balanced-versus-unbalanced lines, shielding and ground loops, types of input stages, overload and methods of padding and, finally, considerations of polarity.

LINE LOSSES

In professional work, *low impedance* microphones are always used. Their source impedance is usually less than 200 ohms, and they are normally loaded by the preamplifier input with about 3000 ohms. Many manufacturers specify that the minimum load impedance be no less than five times the nominal microphone source impedance. For normal cable lengths encountered in the studio (10 meters or so), we need not be concerned with line losses. Good quality microphone cable is usually made up of a pair of 24-gauge (AWG) stranded copper wire placed in a braided shield. The resistance per-unit-length of the copper wire is 0.08 ohms-per-meter, and the capacitance between the two conductors is typically about 100 picofarads-per-meter. Figure 13-1 shows these values in the microphone cable circuit. Obviously, we can ignore the effect of the cable resistance and consider only its capacitance. At 1 kHz, the effect of the 1000 pf-capacitance between the leads is negligible; it amounts to a reactance of −j160K ohms across the resistive load of 3000 ohms. At 20 kHz, the reactance of the capacitance will be −j8000 ohms across the resistive load of 3000 ohms. The net impedance of this combination will be 2868 ohms, as compared with the nominal value of 3000 ohms. The departure is still small enough to be negligible.

Now let us consider the effect of a cable of 60 meters, or about 200

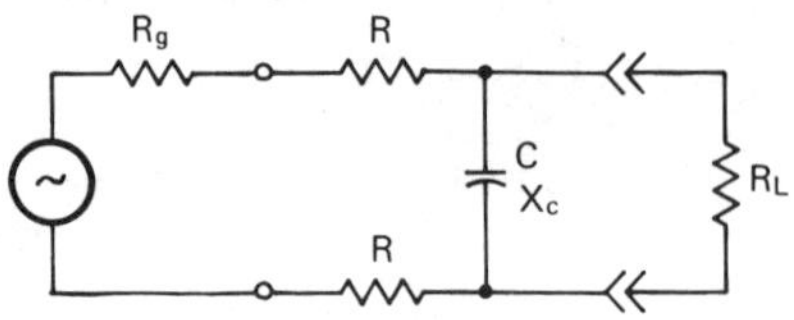

LEGEND:
Rg = Source impedance = 200Ω typical
R = Cable resistance = 0
RL = Load resistance = 3000Ω typical

FREQUENCY	CABLE LENGTH	CABLE CAPACITANCE	REACTANCE	NET IMPEDANCE
1 kHz 20 kHz	10 m	1000 pF	−j160,000Ω −j8000Ω	≈3000Ω 2868Ω
1 kHz 20 kHz	60 m	6000 pF	−j26,500Ω −j1326Ω	≈3000Ω 1213Ω

(A)

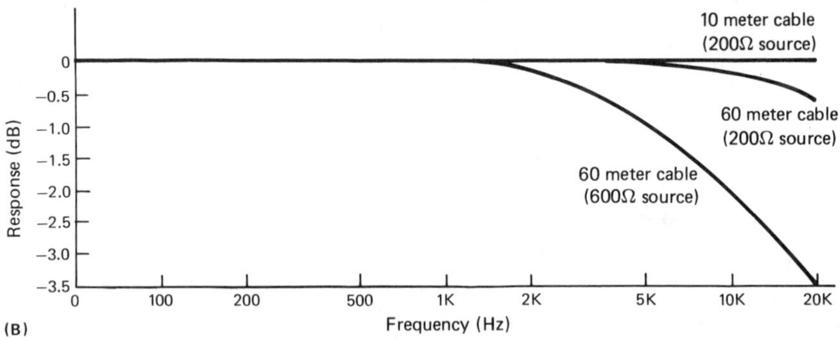

(B)

Figure 13-1. Microphone cable losses. **(A)** In a long cable length, the cable capacitance high frequencies are attenuated, due to the effect of cable capacitance on the system's net impedance. **(B)** Frequency response for various cables and source impedances.

feet. Again, the resistance of the leads is negligible compared with the 3000-ohm load. At 1 kHz, the reactance of the 6000 pf-capacitance will be −j26,500 ohms across the load resistance of 2000 ohms; a negligible amount. However, at 20 kHz, the effect of the cable capacitance will produce a reactance of −j1326 ohms across the resistance of 3000 ohms. The net value of impedance of this combination is 1213 ohms, and the effect of this is to reduce the signal level at 20 kHz by about 1.3 dB.

The reason why the wide swing in load impedance with respect to frequency has a negligible effect on frequency response is that the source impedance of the microphone is much lower than the load impedance, in this case by a factor greater than 10-to-1. If the source impedance were higher, say, 600 ohms, then the effect of the 60-meter cable would reduce the response at 20 kHz by almost 3 dB.

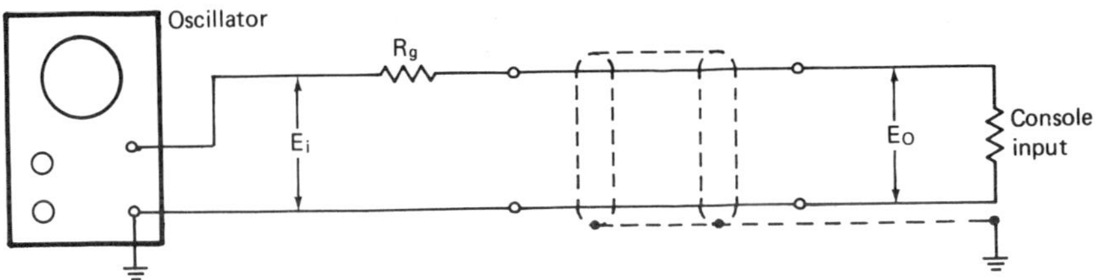

Figure 13-2. Measuring microphone cable losses, by observing E_O as a function of frequency.

Figure 13-1B shows the frequency losses for the several cases we have just discussed.

Clearly, when long cable runs are planned, only low impedance microphones should be used. In general, losses due to cable capacitance in normal recording or sound reinforcement work will be insignificant. However, when there is any doubt, the test setup shown in Figure 13-2 will isolate the problem. While holding the oscillator output voltage E_i, constant, observe the system output vòltage, E_O, as a function of frequency. R_g is the nominal value of source impedance for the microphone intended to be used with the system.

BALANCED-VERSUS-UNBALANCED LINES

In all professional work, balanced microphone lines are used. The virtue of a balanced line is its relative immunity to induced noises. As shown in Figure 13-3A, an interfering noise source induces equal

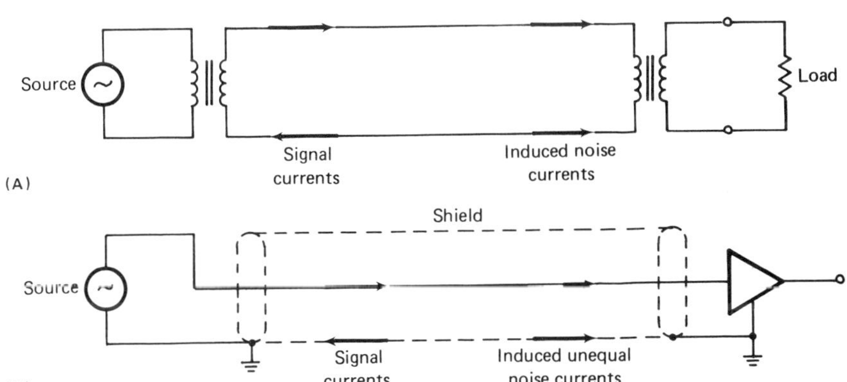

Figure 13-3. Balanced versus unbalanced lines. In the balanced line **(A)**, the equal noise currents induced in both conductors are cancelled. In the unbalanced line **(B)**, the unequal noise currents are not cancelled.

currents in the two conductors; these two currents are of equal amplitude but opposite polarity, and they effectively cancel at the load.

In the case of an unbalanced line, as shown at Figure 13-3B, most of the noise induction will take place in the shield because it is more accessible to the noise source than is the inner conductor. Thus, different noise currents will be induced in the shield and inner conductor, and there will not be total cancellation at the load.

In a balanced system, the cable shield is normally connected to the microphone case and chassis ground of the preamplifier, as shown in Figure 13-4. As a rule, the signal leads are independent of the shield and are "floating" with respect to chassis ground. If the input transformer is center-tapped and connected to ground, the circuit is referred to as *balanced grounded.*

A *ground loop* occurs when elements in a system are connected in such a way that there is continuity in the ground path. If a noise source is located within the loop, then a noise signal may be induced in the loop. The solution to the problem is to break the continuity of the path at some point. A typical example is shown in Figure 13-5.

Ground loop problems are not usually encountered in balanced systems. They are more commonly encountered in unbalanced or *single-ended* systems where the various components may be inadvertently grounded through the chassis of an equipment rack.

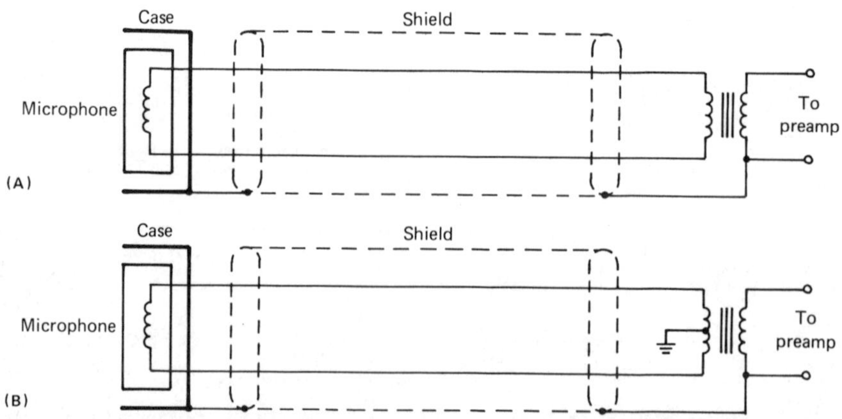

Figure 13-4. Balanced lines. **(A)** A balanced floating line; **(B)** a balanced grounded line.

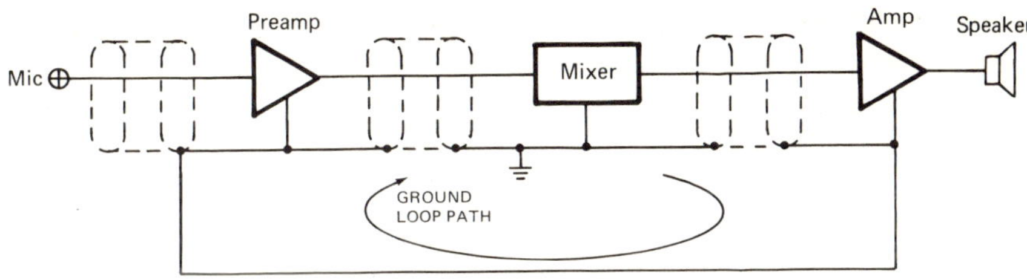

Figure 13-5. A ground loop is created when there is continuity in the ground path.

RADIO FREQUENCY INTERFERENCE (RFI) SHIELDING

The purpose of the shield in a balanced line is to reduce the effects of radio frequency interference. While these signals are well beyond the range of hearing, they can be picked up by cables, acting as antennas. These signals can further be *rectified*, or detected, by unbalanced input stages, with consequent interference at normal audio frequencies. The worst offenders here are television and citizen's band transmissions. Problems can also occur with the various electronically-controlled lighting systems used at rock concerts. Due to the hasty way these productions are often set up, it is easy for microphone cables to be intermixed with lighting power cords.

Good grounding practice calls for a common grounding point, as shown in Figure 13-6. Here, balanced lines are used throughout, except at the loudspeaker output, where signal levels are high enough to ensure immunity from interference.

TYPICAL MICROPHONE INPUT RELATIONSHIPS

Microphone input circuits must be capable of accepting a wide range of input levels without compromise in either distortion or noise performance. Most problems encountered in the field arise out of an improper assessment of gain structure at the input. A comparison of microphones in use today shows that they can vary in sensitivity by as much as 20-25 dB. Further, the peak levels of sound to be picked up may vary from small instrumental groups at, say, 95 dB-SPL up to close rock pickups on the order of 120-125 dB-SPL. Looking at the worst case, peak microphone output levels could vary by as much as 50 dB!

Both microphones and properly-designed input stages are capable of a wider dynamic range than any normal music source. Therefore, there is little excuse for the intrusion of either noise or distortion at the input of a recording system over that which may be inherent in the circuitry itself.

Figure 13-7 outlines several problems. A typical high-quality capacitor microphone may have an effective dynamic range as shown; it is limited at low levels by the input noise of its preamplifier, expressed here as an equivalent sound pressure level weighted for the ear's relative lack of low frequency sensitivity at low levels. The rating of 18 dBA means that the self-noise of the microphone is equivalent to

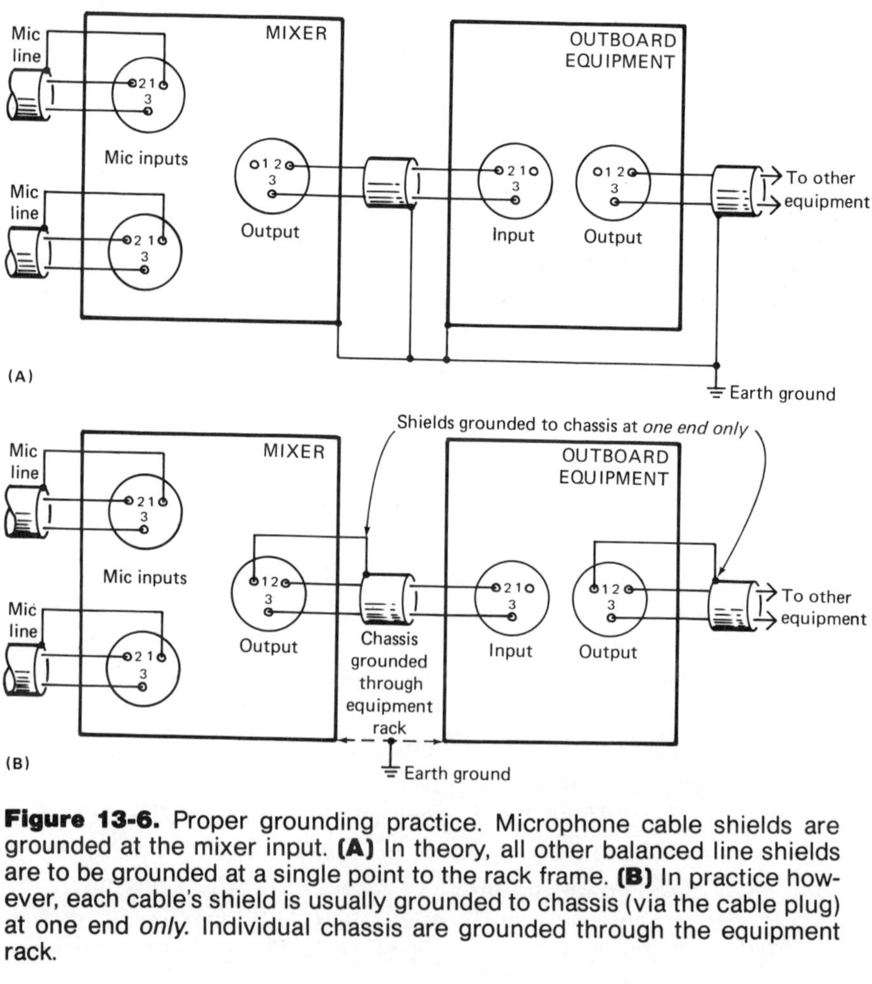

Figure 13-6. Proper grounding practice. Microphone cable shields are grounded at the mixer input. **(A)** In theory, all other balanced line shields are to be grounded at a single point to the rack frame. **(B)** In practice however, each cable's shield is usually grounded to chassis (via the cable plug) at one end *only*. Individual chassis are grounded through the equipment rack.

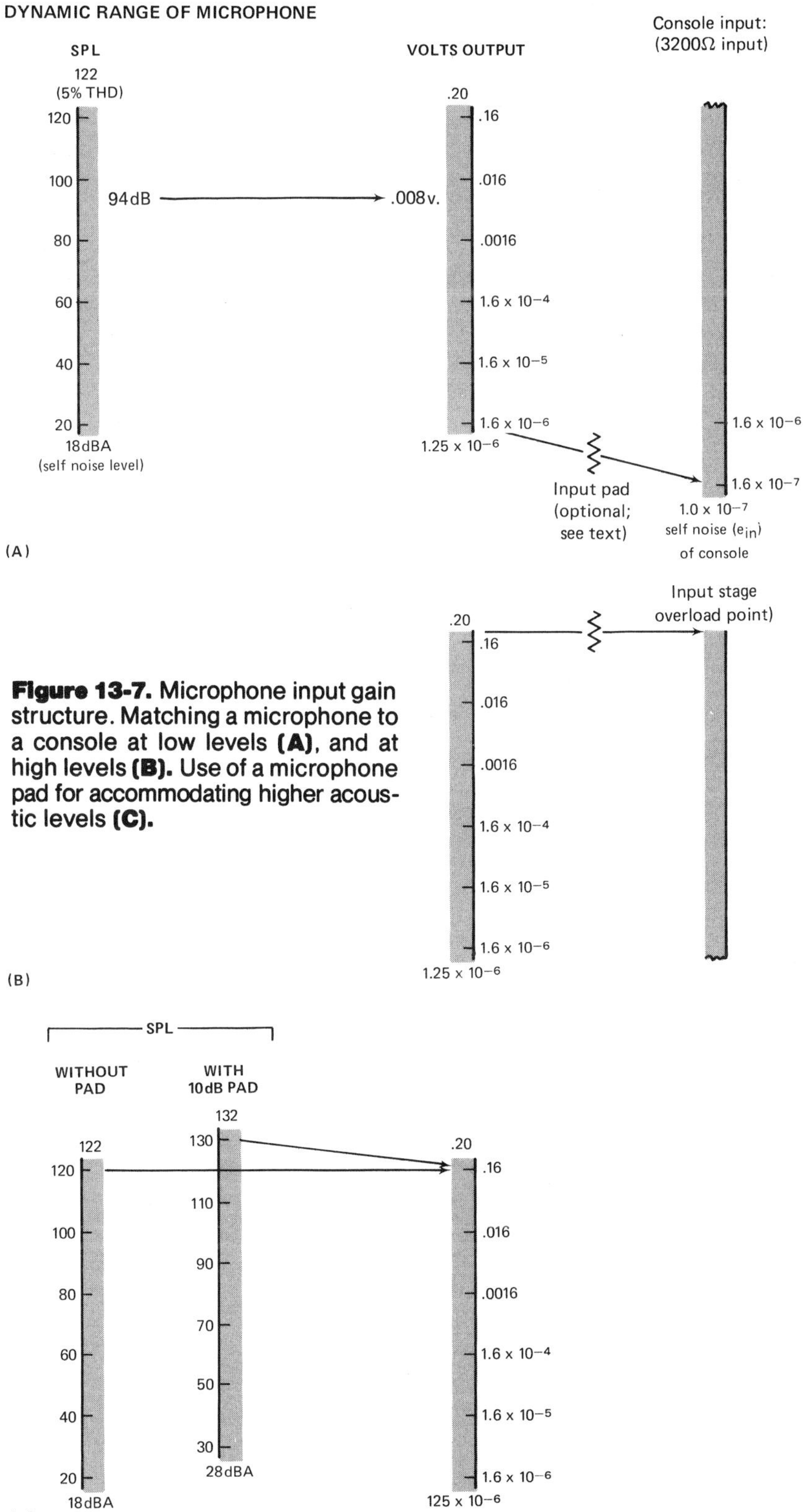

Figure 13-7. Microphone input gain structure. Matching a microphone to a console at low levels **(A)**, and at high levels **(B)**. Use of a microphone pad for accommodating higher acoustic levels **(C)**.

placing it in a sound field which would register a level of 18 dB on a sound level meter through its A-weighting network.

At high levels, a total harmonic distortion (THD) rating of 0.5 percent is arbitrarily established as the upper limit; in this case, 122 dB-SPL. The effective dynamic range, some 104 dB, is greater than any professional recording system operating without auxiliary noise reduction—and it is far greater than any musical program which will be encountered. Furthermore, it is greater than most professional monitoring systems operating in quiet environments are capable of handling.

The microphone's sensitivity is given as 0.008 volts/pascal, or 0.008 volts output in a sound field of 94 dB-SPL. The range of output voltages will then be 1.25 microvolts at the 18-dBA level up to 0.2 volt at the 122 dB level. For lowest noise contribution from the console's input circuitry, gains should be set so that the self-noise of the microphone is well above (at least 10 dB) that of the console's input stages. Therefore, the noise level carried through the recording chain will be that of the microphone *not degraded* by the noise inherent in the console. This general case is shown in Figure 13-7A. Here, we note that the self-noise level of the microphone is some 22 dB above that of the console input (1.25×10^{-6} volts as compared to 1.0×10^{-7} volts). While the microphone could work directly into the console for recording low to normal level program, 12 dB of attenuation could be added ahead of the input stage with no degradation in the microphone's performance on quiet program material.

At the other extreme, the very high sound levels associated with close pickups in pop and rock recording require that the microphone output be adjusted so that it does not overload the input stage of the console. Ideally, the microphone should overload before the input stage does for widest overall dynamic range. This concern will invariably call for *padding,* or attenuating, the microphone's output prior to the input stages, as shown at Figure 13-7B. In this case, the degree of padding may be substantial, pushing the self-noise level of the microphone well below that of the console's input stage.

For extremely high levels, some capacitor microphones allow for padding ahead of the internal preamp so that the limits of the microphone's dynamic range are shifted upward, typically by 10 dB. In Figure 13-7C, the 0.5 percent THD level has been shifted up to 132 dB-SPL and the self-noise level to an equivalent 28 dBA.

In actual recording operations, the dynamic range of the microphone/console system will be much greater than that of the subsequent

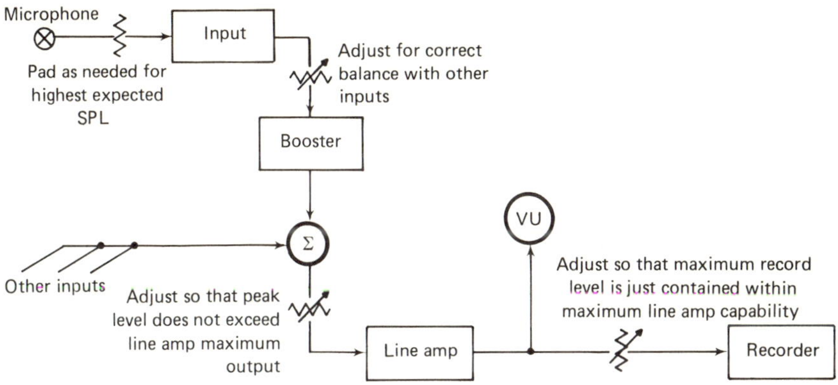

Figure 13-8. Setting levels within the signal chain.

recorder, and the critical gain setting will be that which presents undistorted signal out of the microphone/console system to the recorder at its upper performance limits. In other words, the recorder should overload before the microphone-console combination does. This ideal overall level relationship is shown in Figure 13-8.

The most critical element in a microphone input circuit is often the transformer. Good ones are not cheap. The overall dynamic range of an input device, be it a vacuum tube or a transistor, is normally not much in excess of 100 or 105 dB, and a transformer which will cover this range is a superior one. As we have seen, padding is an important element in matching gains and levels, and it normally takes place ahead of the input transformer. For a balanced input, the pad must also be balanced, and the "H" configuration shown in Figure 13-9A is recommended since it preserves proper impedance relationships over its range of loss settings. In this example, we show resistor values for several values of loss for both 200-ohm and 50-ohm microphones working into a 3000-ohm input. (No values are given for a 25-dB loss with a 50-ohm microphone, since this pad cannot be realized while maintaining matching impedances.)

It is not likely that an engineer will encounter a field situation in which he will have to design and construct a specific loss pad, but it is good to have access to a handbook such as Tremaine's *Audio Cyclopedia* for its information on pads and attenuators.

Where the input transformer is capable of handling a wide range of levels, the circuit arrangement shown in Figure 13-9B is useful. An operational amplifier is used, and negative feedback restrains the input voltage from overloading the circuitry. Because the gain of the

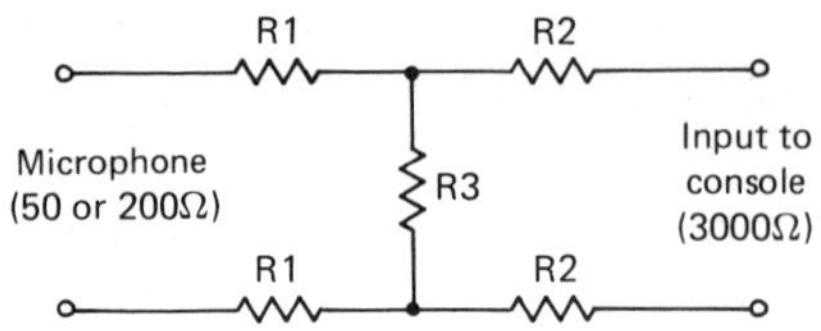

	50Ω			200Ω		
Loss	R1	R2	R3	R1	R2	R3
25dB				1400	12	175
35dB	1500	0	50	1500	75	50
50dB	1500	20	10	1500	95	10

(A)

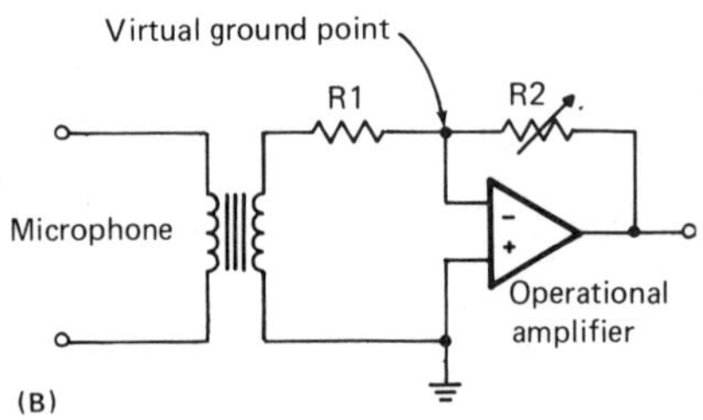

(B)

Figure 13-9. Padding at the input to the console. Passive **(A)** and active **(B)** attenuation of the microphone signal. **(C)** A balanced input, using a differential amplifier.

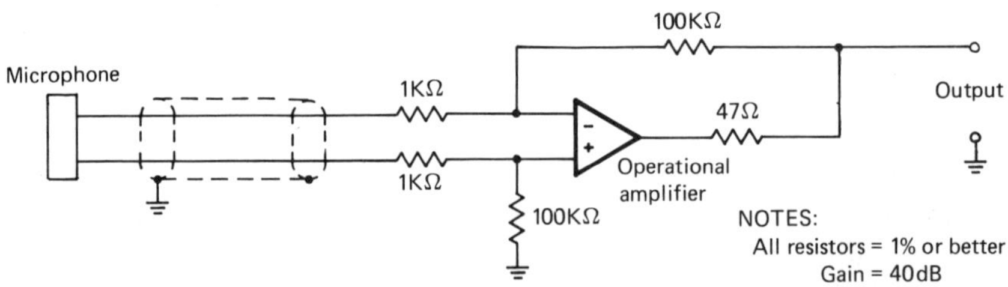

(C)

operational amplifier is quite high (typically about 100-110 dB), the junction between R_1 and R_2 is maintained for all input conditions at a level very close to zero volts; thus the term *virtual ground point.* Under these conditions, any normal input voltage presented to the operational amplifier by the input transformer will not be able to overload it.

Many of the advantages of an input transformer may be met with the *differential input* circuit shown in Figure 13-9C. This provides the balance necessary to result in the cancellation of induced noise that we would routinely expect of a good input transformer. Component values must be carefully matched, and electrical drift is often a problem.

The careful engineer will find out, either by experience or through perusal of manufacturer's specifications, just what the input overload conditions are for each setting of the padding network ahead of the console's input transformer. He will also note the output voltages for the various microphones he is using and relate these to their corresponding sound pressure levels at the microphones.

It helps to have all the microphones of a given type set for approximately the same sensitivity. Because of their internal amplifiers, capacitor microphones can be designed for a wide range of outputs. Typically, they are in the range of 0.5 to 1.5 millvolts open-circuit output for a sound pressure level of 74 dB. Dynamic microphones are typically some 20 dB lower in open-circuit voltage. Values of 0.05 to 0.1 millivolts are common.

Well-designed input stages are always at least five times the nominal impedance of the microphone (some are ten times higher), and this ensures that the microphone will not be adversely loaded down by the input circuit. Figure 13-10 shows what may happen when a microphone is improperly loaded. When the load is equal to the nominal microphone impedance, the level will drop 6 dB. If the microphone's impedance is fairly constant, as in the case of an omni-directional dynamic microphone or a capacitor microphone with its associated preamp, the response will remain smooth. However, in the case of a dynamic cardioid microphone or a ribbon microphone, the deterioration in response will be significant due to the high-impedance peak in the low-frequency range of these microphones, a consequence of their undamped mass-controlled design.

MICROPHONE POLARITY

Polarity is concerned with the sense of the microphone's output voltage relative to pressure maxima and minima at the diaphragm. A basic operating rule in any studio is that *all* microphones be wired for the same polarity. The normal situation is shown in Figure 13-11A. Here, a positive pressure at the microphone results in a positive-going electrical signal at terminal 2 of the output connector. All microphones could just as easily be connected so that pins 2 and 3 were reversed. What is important is that they all be set the same way. If this is not the case, out-of-phase signal conditions could result in inadvertant signal cancellation as well as improper phantom imaging in stereo.

Figure 13-11B shows a simple method for ensuring that all microphones are of the same relative polarity. One by one, the should all be compared with a given microphone which, for the purpose, is assumed to be the reference standard.

Polarity checkers can be used to ensure proper absolute polarity, as shown in Figure 13-11C. These are devices which produce a

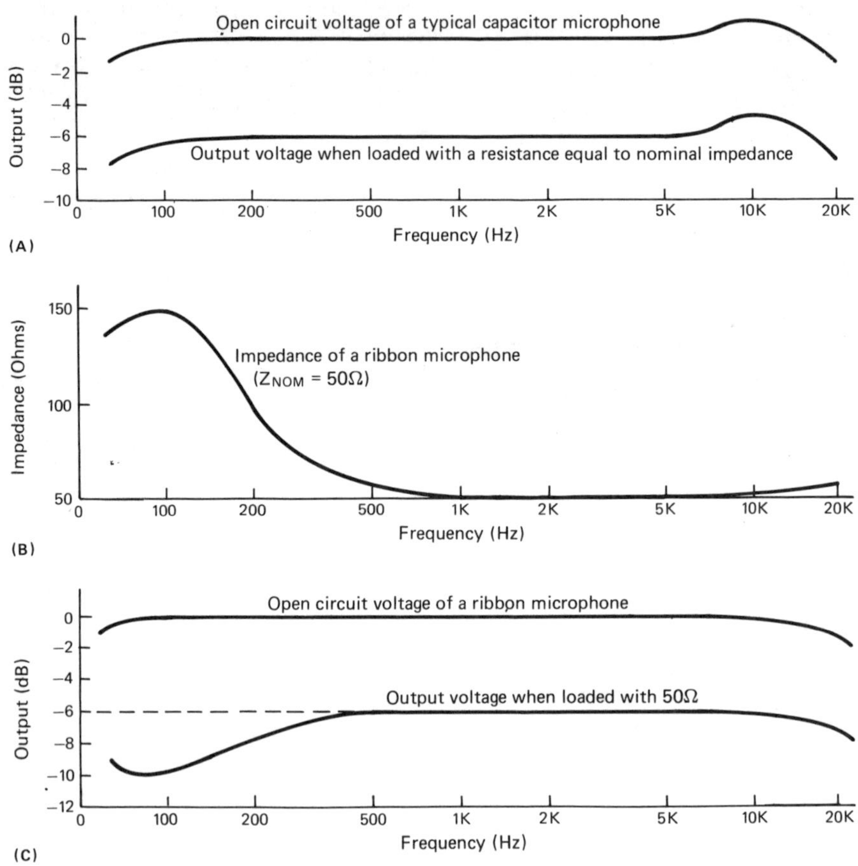

Figure 13-10. Improper loading of a microphone. **(A)** The effect of a matching load on a capacitor microphone; **(B)** the impedance curve of a ribbon microphone; **(C)** the effect of a matching load on a ribbon microphone.

positive pressure pulse at the microphone and indicate, usually via a pair of lights, whether the microphone is properly poled or not.

Other extensions of the notion of relative polarity have to do with inverting and non-inverting amplifiers. For example, a signal processor with an inverting output may be patched into one of two stereo microphones, as shown in Figure 13-12A. The result is an out-of-phase stereo condition which may be corrected simply by inverting one side. However, if the signal path is further complicated by adding a center split signal, such as an in-phase bass line, then the situation simply *cannot* be corrected. Either the bass or the stereo input can be in phase—but not both at the same time. The situation can be

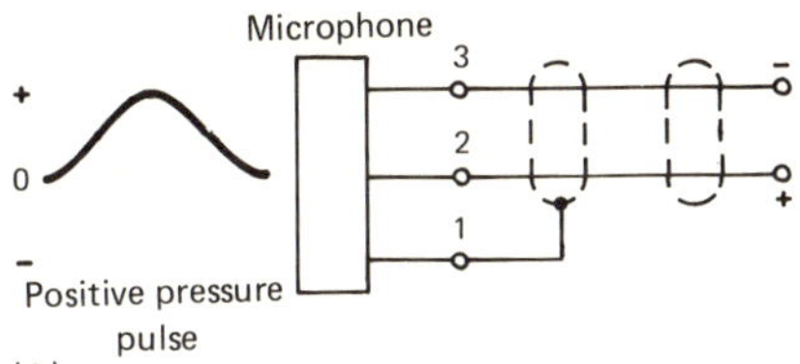

(A)

Figure 13-11. Microphone polarity. **(A)** A positive-pressure pulse should produce a positive voltage at terminal +2 **(B)** comparing an unknown microphone with a standard; **(C)** determining microphone polarity with a polarity checker; **(D)** a commercially-available polarity checker *(Gotham Audio Corp. photo).*

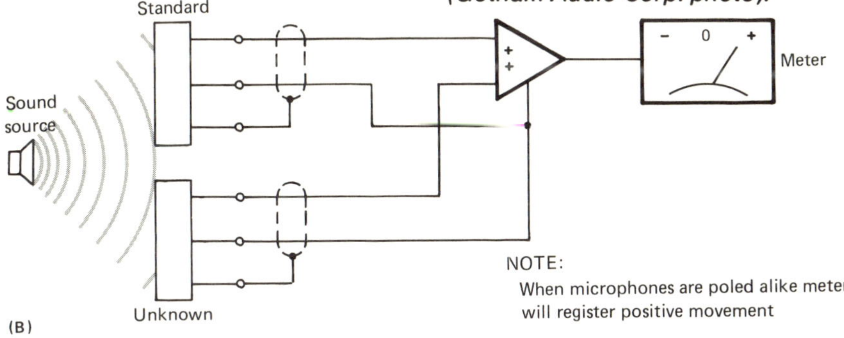

(B)

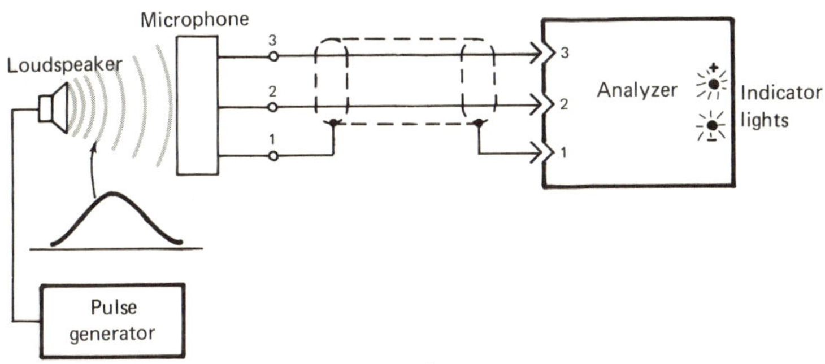

(C)

(D)

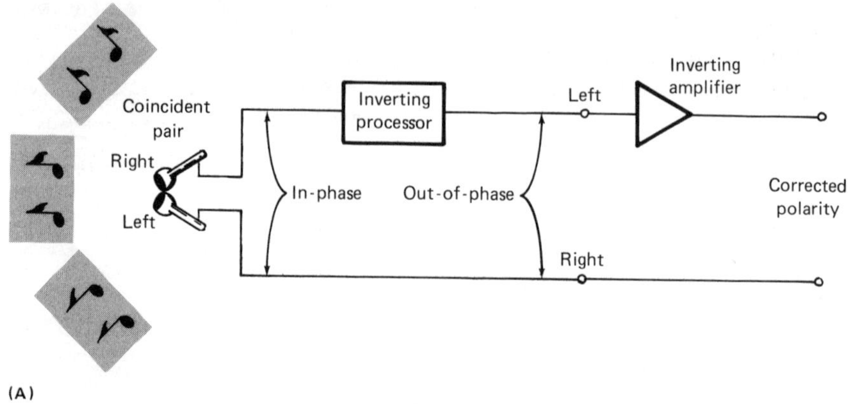

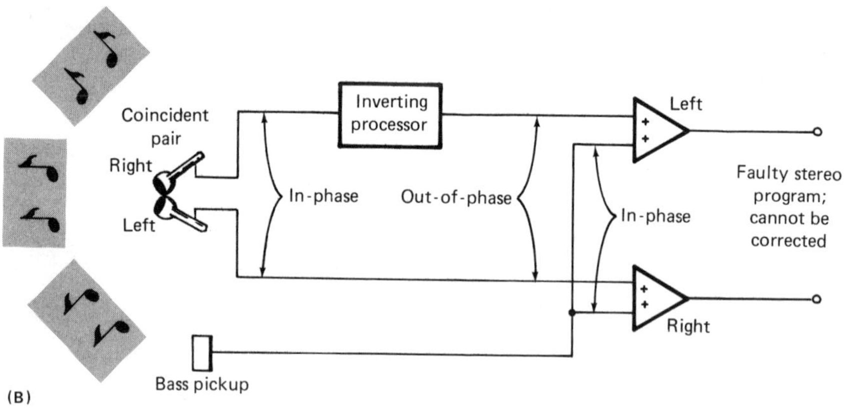

Figure 13-12. Polarity errors in the studio. **(A)** The inverting amplifier corrects the polarity reversal; **(B)** once it is mixed with other signal sources, the polarity reversal cannot be corrected.

corrected only by inserting an *additional* polarity reversal in the line where the inverting output was inserted.

Note also the symmetrical polar pattern of the bi-directional microphone. Which side produces the proper output for a positive input pulse? Some manufacturers indicate by a small "+" sign that axis which will match up with all the other properly-poled microphones. In the other cases it can be assumed that the axis on that side of the microphone bearing the manufacturer's logo produces a correctly-poled signal.

Further considerations of polarity lead to the notion of *absolute polarity*. This is the concern that all of the elements in the complex recording, processing and playback chain have polarity conventions which are strictly adhered to, such that positive-going waveforms sensed by the microphones in the studio will result in positive-going waveforms sensed by the listener in his normal home environment. As strange as it seems, the industry is by no means in agreement as to how the various media and transducers should be poled, and a consensus may be some time away. Meanwhile, there is mounting evidence at the audiophile level that these are important considerations.

CHAPTER XIV

Microphone Accessories

INTRODUCTION

Microphones are rarely used without accessories. The most common accessories are those used in mounting the microphones, and they consist of the following: table mounts, floor stands and booms, special mounts for stereo pickup and permanent mounting of microphones. Shock mounts and wind screens are also important accessories in situations where the microphone must be isolated from adverse environmental conditions. Certain electrical problems, most notably padding and impedance matching, may be solved through the use of accessory in-line pads or transformers.

DESK MICROPHONE MOUNTS

Figure 14-1 shows a variety of table of desk-top microphone mounts. The common desk stand is rapidly being replaced by the flush-mounted form for general use in panel discussions in relatively quiet environments. The grip-to-talk model is useful for paging applications in airports, railroad stations, and wherever else high noise levels prevail and/or where operators can spare only one hand to actuate the paging system. As a rule, the microphone is an integral part of the grip-to-talk mount.

The flush mounted form shown in Figure 14-1B is often referred to as a "mike-mouse," because of its obvious appearance, and is a very economical and handy solution to the problems of physical clumsiness and acoustical cancellation associated with the traditional desk stand. The user should bear in mind that the flush-mounted microphone may often have to be placed a good 20-to-30 centimeters farther away from the talker than the typical desk stand. Also, the flush mount may be somewhat more susceptible to shocks and thumps against the table or desk top than will a standard microphone on a

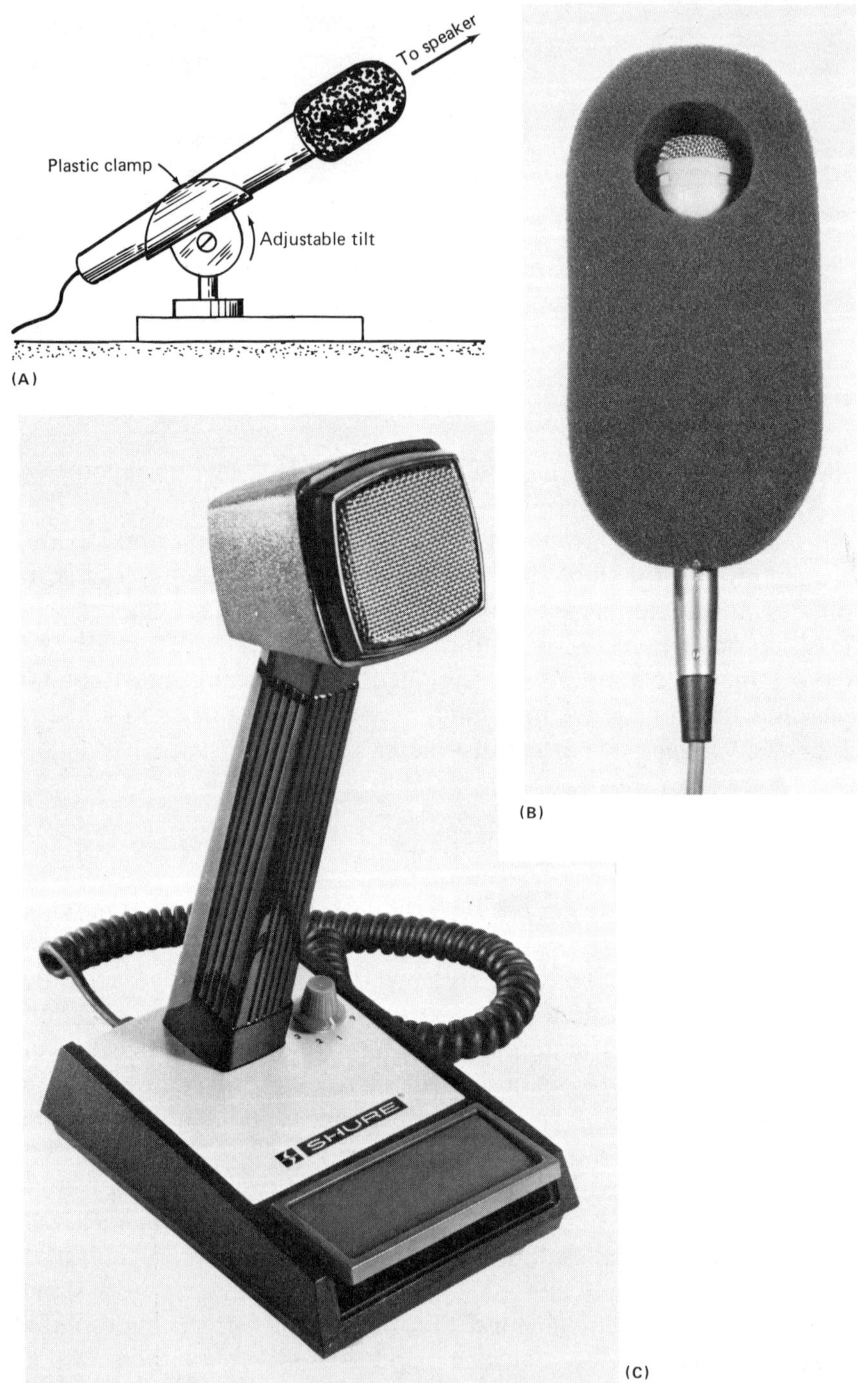

Figure 14-1. Typical desk microphone mounts. **(A)** A desk stand; **(B)** a foam flush mount *(Electro-Voice photo)*; **(C)** a desk stand with hand-actuated switch *(Shure Bros. photo)*.

desk stand. Obviously, an omni-directional microphone should be used with a flush mount.

FLOOR STANDS

Simple floor stands may have circular bases or tripod bases, as shown in Figure 14-2. More expensive models will have casters for ease in rolling around. All floor stands should have telescoping vertical sections for height adjustments and large stands may reach heights of five to six meters. A good microphone stand must have a fairly massive base in order to be dimensionally stable under normal working conditions.

MICROPHONE BOOMS

A microphone boom is a counterbalanced swivel assembly mounted on a stand. Both the stand and the boom can be extended telescopically, and the angle between the boom and the stand can be adjusted. Booms range from small extensions of 1 to 1.5 meters for small microphone stands, up to large assemblies capable of reaching distances of 3 to 4 meters.

The smaller booms are essential in the studio for close microphone placement in hard-to-reach places. For example, the closely-placed microphones used in picking up the elements of a drum set will invariably require booms, because the floor space directly below the microphone is occupied by the instruments themselves. Obviously, the overhead microphones used in picking up a drum set will be on booms as well. The larger booms are most useful in orchestral pickup, where their flexibility in overhead manipulation is appreciated.

A very special kind of microphone boom technique evolved out of the earlier days of motion picture recording. The desire to record dialogue on the set required a boom assembly which could be moved quietly and quickly so that the microphone would not be seen by the camera. Such booms had telescoping sections which could be smoothly extended by the operator, and the microphone mount itself could be turned by the operator to follow the action. The entire assembly was on wheels so that it could be moved as required.

Figure 14-3 shows typical types of booms.

Figure 14-2. Floor microphone stands *(Atlas sound photos).*

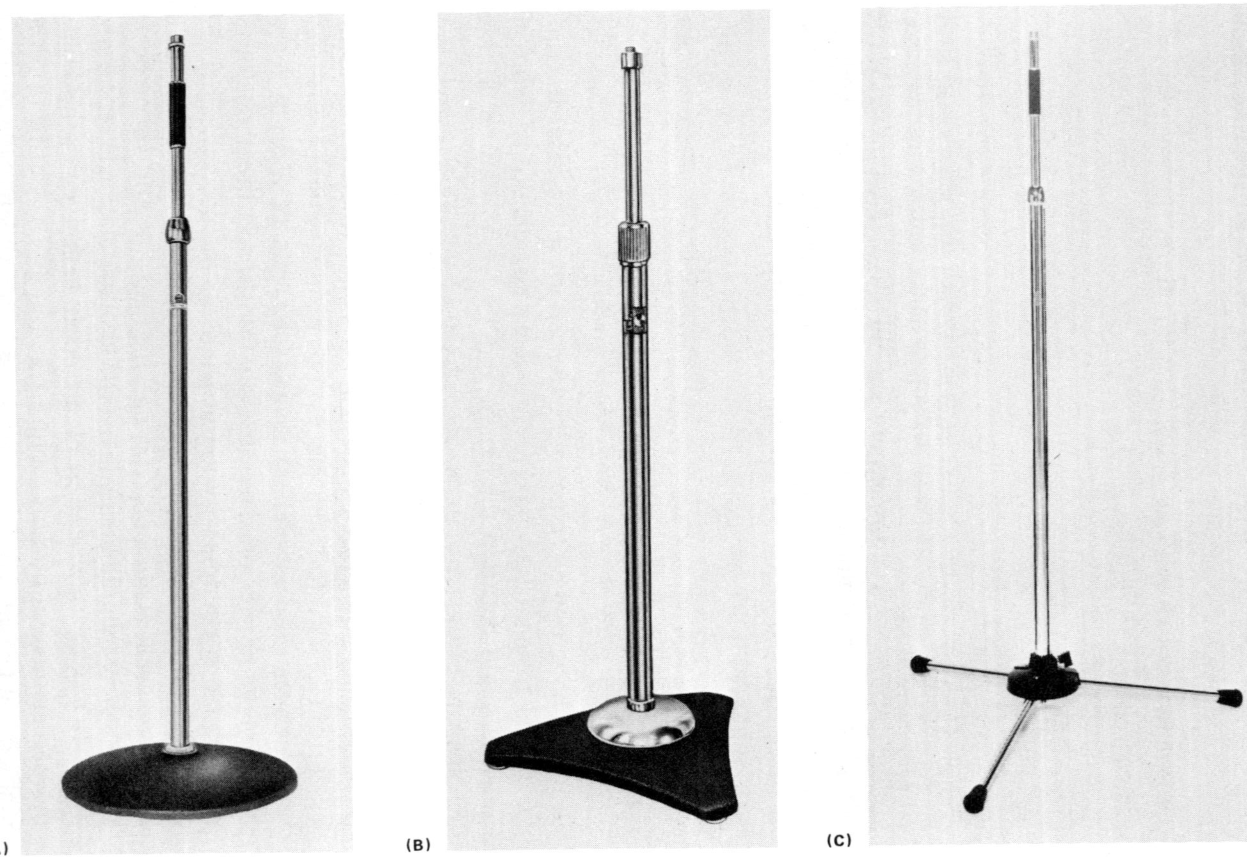

(A) (B) (C)

Figure 14-3. Microphone boom stands.

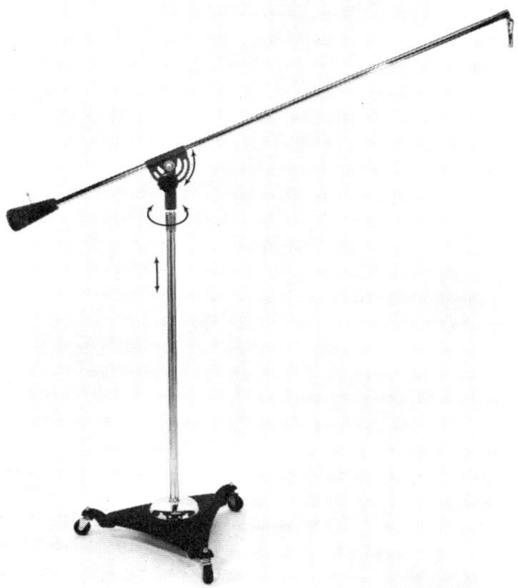

(A)

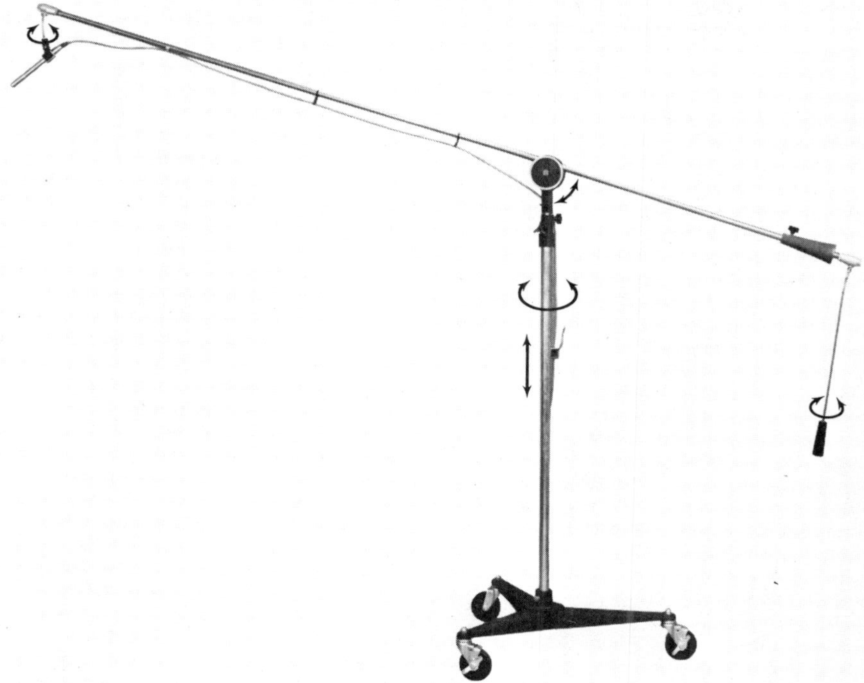

(B)

MICROPHONE MOUNTS

Microphones on Stands and Booms

A microphone on a stand has little mounting flexibility beyond tilting. Figure 14-4A shows typical tilting assemblies for both axial and radial responsive microphones. For boom use, more flexibility may be needed. Figure 14-4B shows typical mounts which allow the microphone's tilt and rotation to be adjusted independently of the boom's vertical angle.

Special Mounts for Stereo Pickup

The increasing popularity of coincident and quasi-coincident microphone techniques has required the development of special mounting hardware for maintaining precise angles and spacing between micro-

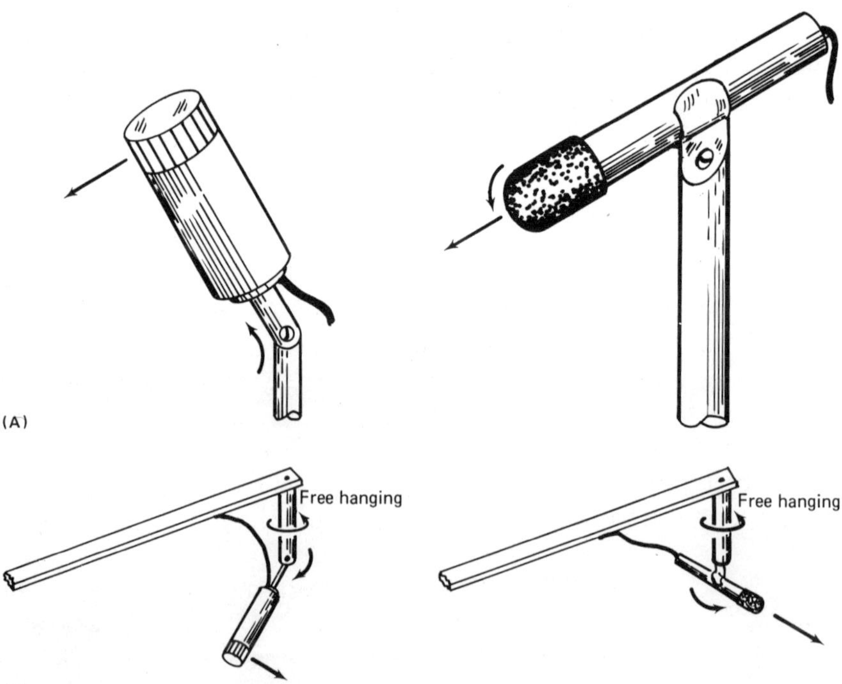

(A)

(B)

Figure 14-4. Microphone mounts for stands and booms. Typical tilting assemblies for stand **(A)** and boom **(B)** applications.

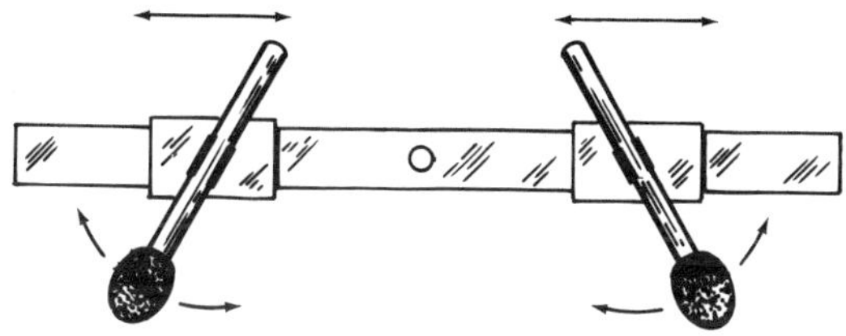

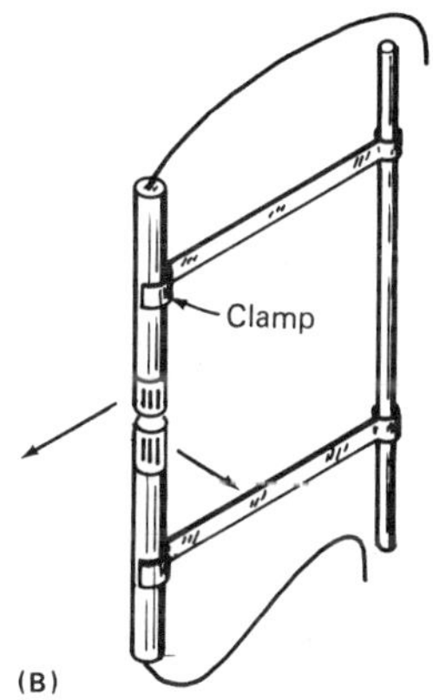

Figure 14-5. Special mounts for stereo pickups suitable for **(A)** ORTF and NOS techniques; **(B)** for coincident-pair recording.

phones. Figure 14-5 shows a mount useful for both ORTF and NOS stereo techniques, and an arrangement that is convenient for mounting a pair of radial-responsive microphones end-to-end for conincident stereo recording.

It is easier to use this kind of hardware for quasi-coincident pickup than to use a pair of microphones each on its own stand. With the latter approach, precise distances and angles are hard to maintain, and even slight shifts of position between microphones will cause significant changes in the stereo perspective.

Permanent Mounting of the Microphones

In many concert halls, especially those where broadcasts originate, some kind of permanent mounting of microphones will be required. Obviously, stands and booms are out of the question if an audience is present. The usual solution is to employ guy wires to stabilize the microphone assembly's position relative to the orchestra. No less than three guy wires will be required, and one of these can be the microphone cable assembly, if proper attention is given to strain relief.

Figure 14-6 illustrates the flexibility in adjusting the position of the microphone array relative to the orchestra by changing the length of the guy wires. However, it is not easy to precisely control the tilt angle of the array relative to the assemble with only three guy wires. For maximum flexibility, a fourth guy wire mounted directly to a lever on the microphone bar will offer the required tilt flexibility.

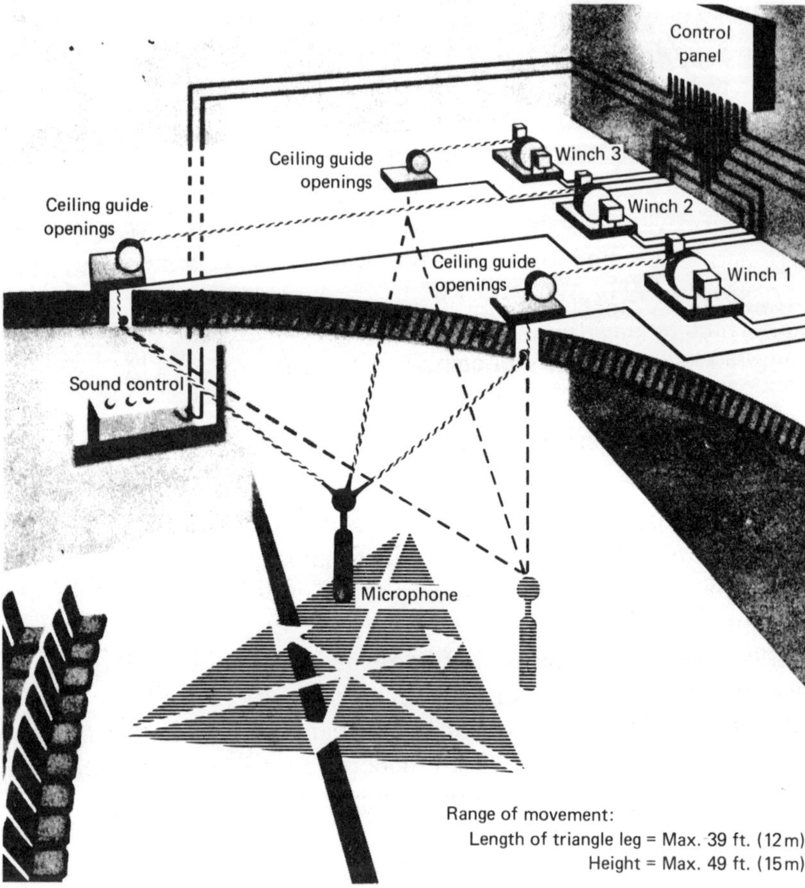

Figure 14-6. A microphone winch system. *(Gotham Audio Corp. photo.)*

As a general recommendation, we advise the user to note very carefully the wide variety of hardware used in mounting accessories. Not all are compatible and nothing is more frustrating than a field expedient which does not work because the hardware does not mate properly. These are details which must be sorted out long before the occasion of their need.

Shock Mounting the Microphone

In many recording environments, the floor can transmit significant low-frequency vibrations to the microphone stand. If these vibrations are not properly isolated, the result can be sub-sonic and audible low-frequency noises in the program. Shock mounts have been developed for this purpose, and typical examples are shown in Figure 14-7. Here, we see the use of a single ring of plastic foam isolating the microphone from its holder. This method is relatively inexpensive and is especially useful for smaller microphones. Another method uses elastic pieces to isolate the microphone from a pair of outer rings. It provides for increased isolation at sub-sonic frequencies and is scaled to handle larger microphones.

As an alternative, blocks of foam under each leg of the microphone stand may isolate the stand from the floor. However, once this has been done, moving the microphone becomes something of a chore.

ISOLATING WIND NOISES

Even a slight breeze across a microphone (especially a directional one) will produce bothersome low-frequency noises. The sharp angles and edges of the microphone case cause turbulence in the air flow, and this produces noise. For instrumentation usage where the microphone will be some distance away from the source under observation, the "balloon" arrangement shown at Figure 14-8A is often used. The larger the surface, the better the isolation at the lowest frequencies. The use of a single layer of silk or other suitable thin weave ensures that high-frequency attenuation will be minimized.

For use at closer quarters, foam screens are required. The open-

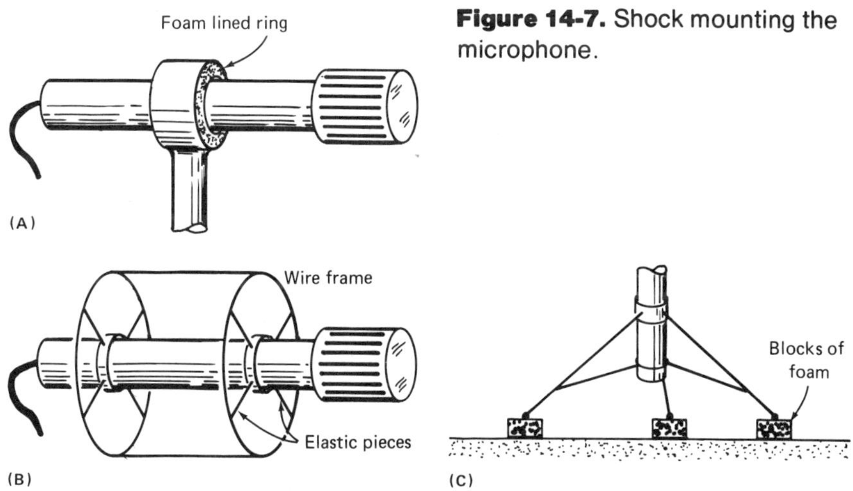

Figure 14-7. Shock mounting the microphone.

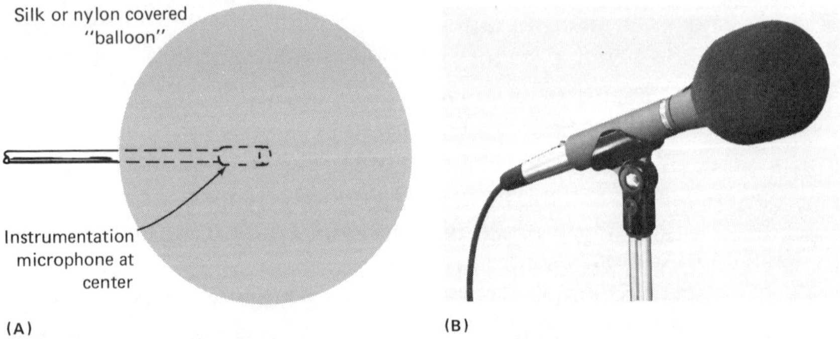

Figure 14-8. Isolating wind noises. **(A)** A silk or nylon covered "balloon"; **(B)** an open-cell foam windscreen *(Electro-Voice photo).*

cell foam construction minimizes high-frequency attenuation. In any event, a slight roll-off at the high end is preferable to wind noise. Incidentally, it should never be necessary to use a wind screen indoors with a hand-held microphone. Proper hand-held microphone technique dictates that the microphones be held slightly to the side of the mouth so that plosive sounds (consonants *p* and *b*) do not impinge on the microphone.

IN-LINE PADS AND TRANSFORMERS

Many companies manufacture a variety of in-line devices for quick field usage. Some of these are shown in Figure 14-9. An in-line transformer is used in a non-professional situation where a low-impedance microphone is to be matched to a high-impedance input.

The attenuation pad is used to attenuate excessive levels prior to the mixer input. A high-pass filter is useful for attenuating low-frequency proximity effects, or in situations where there is no equalization capability in the mixer.

Simple male-to-male and female-to-female adapters may come in handy in those unexpected situations where male input jacks are encountered in a mixer instead of the usual female type. In-line pads and transformers are also available for going from balanced lines to unbalanced inputs. In these cases, conductors one and two are shorted together and grounded at the input.

A number of manufacturers have built into their microphones many of the features normally encountered in external accessories. For example, some modern dynamic directional microphones are optimized for hand-held use through internal shock mounting. These designs minimize case noise due to handling of the microphone. In addition to the commonly encountered speech/music low roll-off switch found in many microphones, some recent designs include the rudiments of in-line equalization (groups of three-position switches offering boost or dip at several frequencies) as well as signal limiting in order to prevent input stage overload.

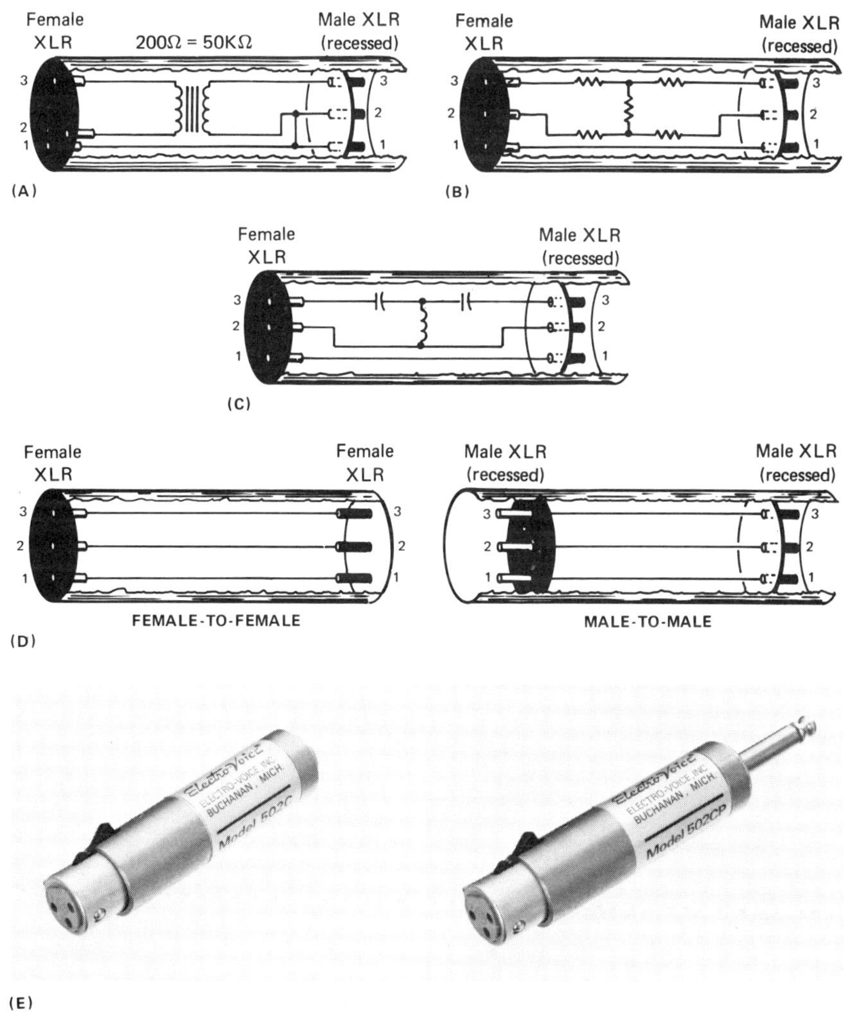

Figure 14-9. In-line microphone accessories. **(A)** transformer; **(B)** attenuation pad; **(C)** equalizer (filter); **(D)** female-to-female and male-to-male adapters; **(E)** typical in-line devices *(Electro-Voice photos).*

Microphone systems are also popular. These are composed of groups of capacitor capsules, each optimized for a specific application and all interchangeable with a common amplifier-polarizing assembly. Typical of the individual microphone elements would be: omni (flat direct incidence), omni (flat random incidence), cardioid (flat), cardioid (low-frequencies rolled off for speech), hyper-cardioid and bi-directional.

Bibliography

ADDITIONAL READING

Author's Note: There have been very few books devoted entirely to microphones. However, microphones have been treated amply in technical papers and through analysis of their roles in recording, broadcasting and sound reinforcement. The bibliography presented here divides the subject into four areas: general and theoretical considerations, stereophonic and quadraphonic recording, studio recording techniques, and sound reinforcement.

In the section on general and theoretical considerations, the microphone is treated somewhat apart from its application. These articles and books deal essentially with basic physical principles and engineering considerations involved in microphone design. The AES (Audio Engineering Society) anthology is highly recommended for the breadth of its coverage, as is the little gem of a book by Boré, which should be in every engineer's library.

The section on stereophonic and quadraphonic recording treats the ways groups of microphones respond, both individually and collectively, to sound fields existing in a performance environment. The classic reference in this important area is the Blumlein patent of 1931. It should be required reading for all aspiring recording engineers.

The section on studio recording may appear to some readers to be rather short. Hardly an issue of a trade journal goes by without an article dealing with studio application, complete with identification of specific microphone models and types favored by the writer or, as is more often the case, the interviewee. These articles are useful if viewed in proper perspective, but they do not deserve "paper" status. Only those papers which adequately cover the reasons behind microphone choice have been included here, and even then, a good bit of culling has been done.

The section on sound reinforcement reflects the professionalism which has come into that field only in the last 10 to 12 years. Not listed here are the valuable proprietary training manuals which have been issued over the years by the major manufacturers of professional sound equipment. They are worth looking for.

For the most part, the references included here are fairly recent; only those earlier references of more than normal significance have been included. Those entries followed by an asterisk are included in the appropriate AES anthologies cited here.

GENERAL AND THEORETICAL

Books and Anthologies:

L. Beranek, *Acoustics* (McGraw-Hill, New York 1954).

G. Boré, "On the Behavior of Directional Microphones," (Gotham Audio Engineering Bulletin, New York).

L. Burroughs, *Microphones: Design and Application*, (Sagamore Publishing, Plainview, New York 1974).

J. Eargle, *Sound Recording* (Van Nostrand Reinhold, New York 1980).

J. Frayne and H. Wolfe, *Elements of Sound Recording*, (John Wiley, New York 1949).

O. Read, *The Recording and Reproduction of Sound*, (Howard W. Sams, Indianapolis, 1952).

A.E. Robertson, *Microphones* (Hayden Book Co., New York, 1963).

H. Tremaine, *Audio Cyclopedia*, (Howard W. Sams, Indianapolis, 1969).

J. Woram, *The Recording Studio Handbook*, (Sagamore Publishing, Plainview, New York 1980).

Various, *Microphones*, (An anthology of articles on microphones from the pages of the *J. Audio Eng. Soc.*, vol. 1 through vol. 27).

Various, *Motion Picture Sound Engineering*, (D. Van Nostrand, New York 1938).

Papers:

F. Bauch, "New High-Grade Condenser Microphones," *J. Audio Eng. Soc.*, vol. 1, no. 3 (1953).*

G. Boré, "Powering Condenser Microphones," *db Magazine*, vol. 4, no. 6 (1970).

J. Dorner, "Input Noise in Microphones Preamplifiers," *db Magazine*, vol. 12, no. 7 (1978).

J. Eargle, "How Capacitor Microphones Produce Cardioid Patterns," *db Magazine* vol. 5, no. 4 (1971).

K. Elliot, "CMOS Microphone Cable Tester," *db Magazine*, vol. 13, no. 12 (1979).

H. Olson, "Directional Microphones," *J. Audio Eng. Soc.*, vol. 15, no. 4 (1967).*

R. Schulein, "Microphone Considerations in Feedback-Prone Environments," *J. Audio Eng. Soc.*, vol. 24, no. 6 (1976).*

R. Schulein, "Electrets and Condenser Microphones," *db Magazine*, vol. 12, no. 7 (1978).

G. Sessler and J. West, "Condenser Microphones with Electret Foil," *J. Audio Eng. Soc.* vol. 12, no. 2 (1964).*

R. Werner, "On Electrical Loading of Microphones," *J. Audio Eng. Soc.*, vol. 3, no. 4).*

STEREOPHONIC AND QUADRAPHONIC RECORDING

Books:

J. Borwick, Sound Recording Practice, (Oxford University Press, London 1980).

J. Eargle, *Sound Recording*, (Van Nostrand Reinhold, New York 1980).

J. Woram, *The Recording Studio Handbook*, (Sagamore Publishing, Plainview, New York 1980).

Papers:

B. Bartlett, "Stereo Microphone Technique," *db Magazine,* vol. 13, no. 12 (1979).

A. Blumlein, "British Patent Specifications 394,325 (Directional Effect in Sound Systems)," *J. Audio Eng. Soc.,* vol. 6, no. 2 (1958).

C. Ceoen, "Comparative Stereophonic Listening Tests," *J. Audio Eng. Soc.,* vol. 20, no. 1 (1972).

H. Clark, G. Dutton, P. Vanderlyn, "The 'Stereosonic' Recording and Reproducing System," *J. Audio Eng. Soc.,* vol. 6, no. 2 (1958).

M. Gerzon. "Practical Periphony: The Reproduction of Full-Sphere Sound," presented at AES Convention, London, 25 February 1980.

C. Huston, "A Quadraphonic Microphone Development," *Recording Engineer / Producer,* vol 1, no. 3 (1970).

T. Lubin, "The Calrec Soundfield Microphone, *Recording Engineer / Producer,* vol. 10, no. 6 (1979).

E. Madsen "The Application of Velocity Microphones to Stereophonic Recording," *J. Audio Eng. Soc.,* vol. 5, no. 2 (1957).

J. Mosely, "Eliminating the Stereo Seat," *J. Audio Eng. Soc.,* vol. 8, no. 1 (1960).

J. Smith, "The Sound Field Microphone," *db Magazine,* vol. 12, no. 7 (1978).

C. Taylor, "The Calrec Sound Field Microphone in Use," *Recording Engineer / Producer,* vol. 10, no. 6 (1979).

R. Wickersham and E. Long, "How to Make Convincing Stereo Recordings," *High Fidelity Magazine* (August 1978).

STUDIO RECORDING TECHNIQUES

Books:

J. Borwick, *Sound Recording Practice,* (Oxford University Press, London 1980).

L. Burroughs, *Microphones: Design and Application,* (Sagamore Publishing, Plainview, New York 1974).

J. Eargle, *Sound Recording*, (Van Nostrand Reinhold, New York 1980).

A. Everest, *Handbook of Multichannel Recording*, (TAB Books, Blue Ridge Summit, Pa., 1975).

R. Runstein, *Modern Recording Techniques*, (Howard W. Sams, Indianapolis, 1974).

J. Woram, *The Recording Studio Handbook*, (Sagamore Publishing, Plainview, New York 1980).

Papers:

E. Long, "The Pressure Recording Process," *db Magazine*, vol. 14, no. 1 (1980).

M. Thorne, "Studio Microphone Technique," *Studio Sound*, vol. 15, no. 7 (1973).

P. White, "Distortion-Measuring Microphones for Music Recording," *db Magazine*, vol. 14, no. 1 (1980).

W. Woszczyk, "Improved Instrument Timbre Through Microphone Placement," *Recording Engineer/Producer*, vol. 10, no. 5 (1979).

SOUND REINFORCEMENT

Books and Anthologies:

D. and C. Davis, *Sound System Engineering*, (Howard W. Sams, Indianapolis, 1975).

Various, *Sound Reinforcement*, (an anthology of articles on sound reinforcement from the pages of the *J. Audio Eng. Soc.*, vol. 1 through vol. 26).

Papers:

C.P. and R.E. Boner, "The Gain of a Sound System," *J. Audio Eng. Soc.*, vol. 17, no. 2 (1969).*

D. Dugan, "Automatic Microphone Mixing," *J. Audio Eng. Soc.*, vol. 23, no. 6 (1975).

E. Jones, "Providing Fold-Back with Out-of-Phase Loudspeakers," *J. Audio Eng. Soc.*, vol. 19, no. 4 (1971).*

M. Rettinger, "Sound Reinforcement Systems in Amphitheatres," *db Magazine*, vol. 14, no. 5 (1980).

R. Schulein, "*In Situ* Measurement and Equalization of Sound Reproducing Systems," *J. Audio Eng. Soc.*, vol. 23, no. 3 (1975).*

Index

A

B

C

D

E

T

U

V

W

NOTES

NOTES

NOTES

NOTES

NOTES

NOTES

NOTES

NOTES